# DIGITAL VIDEO PROCESSING FOR ENGINEERS

# DIGITAL VIDEO PROCESSING FOR ENGINEERS

## A Foundation for Embedded Systems Design

MICHAEL PARKER, SUHEL DHANANI

AMSTERDAM • BOSTON • HEIDELBERG • LONDON
NEW YORK • OXFORD • PARIS • SAN DIEGO
SAN FRANCISCO • SINGAPORE • SYDNEY • TOKYO
Newnes is an Imprint of Elsevier

Newnes is an imprint of Elsevier
The Boulevard, Langford Lane, Kidlington, Oxford OX5 1GB, UK
225 Wyman Street, Waltham, MA 02451, USA

First edition 2013

**Notice**
No responsibility is assumed by the publisher for any injury and/or damage to persons or property
as a matter of products liability, negligence or otherwise, or from any use or operation of any
methods, products, instructions or ideas contained in the material herein. Because of rapid advances
in the medical sciences, in particular, independent verification of diagnoses and drug dosages should
be made

**British Library Cataloguing in Publication Data**
A catalogue record for this book is available from the British Library

**Library of Congress Cataloging-in-Publication Data**
A catalog record for this book is availabe from the Library of Congress
ISBN: 978-0-12-415760-6

For information on all Newnes publications visit
our web site at books.elsevier.com

Working together to grow
libraries in developing countries

www.elsevier.com | www.bookaid.org | www.sabre.org

ELSEVIER     BOOK AID     Sabre Foundation
             International

# CONTENTS

**Chapter 1  Video in the Modern World** ..............................................1

**Chapter 2  Introduction to Video Processing** ...........................5

  **2.1**  Digital Video: Pixels and Resolution..............................5

  **2.2**  Digital Video: Pixels and Bits.........................................6

  **2.3**  Digital Video: Color Spaces ..........................................8

  **2.4**  Video Processing Performance ......................................9

**Chapter 3  Sampling and Aliasing** ...................................11

  Digitizing and Sampling .............................................13

  **3.1**  Nyquist Sampling Rule .................................................16

**Chapter 4  Introduction to Digital Filtering** ..................19

  **4.1**  Median Filtering .............................................................19

  **4.2**  FIR Filters .....................................................................20

  **4.3**  FIR Filter Construction .................................................20

  **4.4**  Computing Frequency Response..................................23

**Chapter 5  Video Scaling** ...................................................29

  **5.1**  Understanding Video Scaling .......................................30

  **5.2**  Implementing Video Scaling .........................................33

  **5.3**  Video Scaling for Different Aspect Ratios....................36

  **5.4**  Conclusion ....................................................................38

**Chapter 6  Video Deinterlacing** .......................................39

  **6.1**  Basic Deinterlacing Techniques ..................................40

  **6.2**  Motion-Adaptive Deinterlacing: The Basics ...............43

  **6.3**  Logic Requirements......................................................45

  **6.4**  Cadence Detection .......................................................47

  **6.5**  Conclusion ....................................................................48

**Chapter 7  Alpha Blending**..................................................**49**
**7.1** Introduction ..................................................49
**7.2** Concept and Math Behind Alpha Blending ..................50
**7.3** Implementing Alpha Blending in Hardware ..................51
**7.4** Creating a Different Background..................51
**7.5** Conclusion..................................................52

**Chapter 8  Sensor Processing for Image Sensors**..................**53**
**8.1** CMOS Sensor Basics..................................................54
**8.2** A Simplistic HW Implementation of Bayer Demosaicing ..................56
**8.3** Sensor Processing in Military Electro-optical Infrared Systems ..................56
**8.4** Conclusion..................................................58

**Chapter 9  Video Interfaces**..................................................**61**
**9.1** SDI..................................................61
**9.2** Display Port ..................................................61
**9.3** HDMI..................................................63
**9.4** DVI..................................................63
**9.5** VGA..................................................64
**9.6** CVBS..................................................64
**9.7** S-Video..................................................64
**9.8** Component Video..................................................65

**Chapter 10  Video Rotation**..................................................**67**
Image Rotation Matrix ..................................................67
**10.1** Interpolation ..................................................68

**Chapter 11  Entropy, Predictive Coding and Quantization**..................**69**
**11.1** Entropy..................................................69
**11.2** Huffman Coding..................................................71
**11.3** Markov Source..................................................72

**11.4** Predictive Coding ................................................................73

**11.5** Differential Encoding ..........................................................74

**11.6** Lossless Compression .........................................................75

**11.7** Quantization........................................................................76

**11.8** Decibels...............................................................................79

**Chapter 12 Frequency Domain Representation** ...........................**83**

**12.1** DFT and IDFT Equations......................................................85

**12.2** Fast Fourier Transform ........................................................93

**12.3** Discrete Cosine Transform ..................................................97

**Chapter 13 Image Compression Fundamentals**............................**103**

**13.1** Baseline JPEG ................................................................... 103

**13.2** DC Scaling ........................................................................ 104

**13.3** Quantization Tables .......................................................... 104

**13.4** Entropy Coding ................................................................. 106

**13.5** JPEG Extensions ............................................................... 108

**Chapter 14 Video Compression Fundamentals** ...........................**111**

**14.1** Block Size........................................................................... 112

**14.2** Motion Estimation.............................................................. 114

**14.3** Frame Processing Order .................................................... 116

**14.4** Compressing I-frames ....................................................... 117

**14.5** Compressing P-frames ...................................................... 118

**14.6** Compressing B-frames....................................................... 119

**14.7** Rate Control and Buffering ................................................ 119

**14.8** Quantization Scale Factor ................................................. 120

**Chapter 15 From MPEG to H.264 Video Compression** .................**125**

**15.1** MPEG-2 ............................................................................. 126

**15.2** H.264 Video Compression Standard.................................. 131

**15.3** Digital Cinema Applications .............................................. 140

## Chapter 16 Video Noise and Compression Artifacts ....................141

**16.1** Salt-and-pepper Noise ................................................. 141

**16.2** Mosquito Noise ......................................................... 143

**16.3** Block Artifacts ....................................................... 144

## Chapter 17 Video Modulation and Transport ...........................147

**17.1** Complex Modulation and Demodulation ......................... 147

**17.2** Modulated Signal Bandwidth ...................................... 150

**17.3** Pulse Shaping Filter ................................................ 152

**17.4** Raised Cosine Filter ................................................ 155

**17.5** Signal Upconversion ................................................ 163

**17.6** Digital Upconversion ............................................... 164

## Chapter 18 Video over IP ...............................................169

**18.1** Basics of Internet Protocol (IP) ................................. 169

**18.2** Encapsulation ........................................................ 171

**18.3** Video Streams ....................................................... 171

**18.4** Transport Protocols ................................................ 172

**18.5** IP Transport ......................................................... 173

**18.6** Video Over Internet Issues ....................................... 175

**18.7** Video Streaming ..................................................... 176

**18.8** Multicast Video ...................................................... 177

**18.9** Video Conferencing ................................................. 178

## Chapter 19 Segmentation and Focus ..................................181

**19.1** Measuring Focus ..................................................... 182

**19.2** Segmentation ........................................................ 185

## Chapter 20 Memory Considerations When Building a Video Processing Design .............................................191

**20.1** The Frame Buffer .................................................... 191

**20.2** Calculating External Memory Bandwidth Required ............ 194

**20.3** Calculating On-Chip Memory ........................................ 199

**20.4** Conclusion ........................................................ 200

**Chapter 21 Debugging FPGA-based Video Systems**......................**201**

**21.1** Timing Analysis ................................................ 201

**21.2** The SystemConsole Debugger ................................. 204

**21.3** Check That Clocks and Resets are Working.................. 205

**21.4** Clocked and Flow Controlled Video Streams ............... 206

**21.5** Debugging Tools ............................................. 206

**21.6** Converting from Clocked to Flow-controlled Video Streams . 208

**21.7** Converting from Flow-controlled to Clocked Video Streams . 209

**21.8** Free-running Streaming Video Interfaces ................. 210

**21.9** Insufficient Memory Bandwidth ............................ 211

**21.10** Check Data Within Stream ................................ 212

**21.11** Summary .................................................... 213

Index ................................................................. 215

# VIDEO IN THE MODERN WORLD

Video began as a purely analog technology. Successive images were captured on film streaming through the camera. The movie was played by using flashes of light to illuminate each frame on the moving film, at rates sufficient to show continual motion. Flicker, however, was easily seen.

An improved system for early broadcast television utilized the luminance (or light intensity) information represented as an analog signal. To transmit an image, the luminance information was sent in successive horizontal scans. Sufficient horizontal scans built up a two-dimensional image. Televisions and monitors used cathode ray guns that shot a stream of electrons to excite a phosphorus-coated screen. The slowly fading phosphorus tended to eliminate flicker. The cathode gun scanned in successive rows, each row just below the previous row, guided by magnetic circuits. This happened so rapidly that images were "painted" at a rate of 25 to 30 frames per second (fps). The luminance signal was used to control the intensity of the electron stream.

A horizontal synchronization signal is used to separate the horizontal scan periods. The horizontal sync period is a short pulse at the end of each scan line. This has to be long enough to allow the electron gun to move back to the left side of the screen, in preparation for the next scan line. Similarly, the vertical synchronization signal occurs at the end of the last or bottom scan, and is used to separate each video frame. The vertical synchronization interval is much longer, and allows the electron gun to move back up from the lower right corner of the screen to the upper left corner, to begin a new frame.

Later, color information in the form of red and blue hues was added, known as chrominance information. This was superimposed on the luminance signal, so that the color system is backwards compatible to the black-and-white system.

Modern video signals are represented, stored and transmitted digitally. Digital representation has opened up all sorts of new usages of video. Digital video processing is of growing importance in a variety of markets such as video surveillance; video conferencing; medical imaging; military imaging

Digital Video Processing for Engineers. http://dx.doi.org/10.1016/B978-0-12-415760-6.00001-5

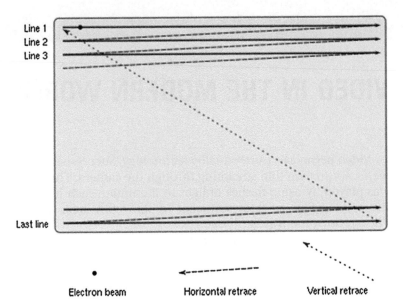

**Figure 1.1.** Video raster scan.

including UAVs (Unmanned Aerial Vehicles), weapons sights and night vision; broadcast; digital cinema; industrial displays and consumer electronics. All these sectors are embarking on a massive decade-long upgrade cycle from standard definition to HD-and higher than HD-resolution video processing. In some cases the old analog video-processing equipment is being replaced by systems using digital techniques.

In many cases, industries that have not traditionally been involved in video processing must now integrate this technology into their products. Examples are rear cameras, entertainment centers, "lane departure" and "head-up" displays in automotives; video data handling in networking servers and routers; sharing and merging of video to provide situational awareness in military systems; surveillance and guidance in military and commercial airborne systems; robotic systems; facial and other features recognition (such as license plates) in security surveillance systems and myriad other applications. This trend is placing new requirements on system designers as well as implementation engineers to understand video technology. This book is designed for those individuals who need to understand basic concepts and applications, so that they can either build their own video systems, or integrate third-party video technology into their products.

Another target audience is those involved in technical marketing and sales and executives from the many industries requiring video technology. Again, the need is to understand basic concepts and applications, without being overwhelmed by the details and implementation complexity.

The market sizes of these new video applications are growing rapidly. For example, here are some publically available projections:

- ABI Research believes that the video surveillance market is poised for explosive growth, which the firm forecasts to expand from revenue of about $13.5 B in 2006 to a remarkable $46 B in 2013. Those figures include cameras, computers and storage, professional services, and hardware infrastructure: everything that goes into an end-to-end security system.
- According to Wainhouse Research, the overall endpoint market for video conferencing will grow from $1.3 B in 2007 to over $4.9 B in 2013. Videoconferencing infrastructure product revenues, including MCUs, gateways, and gate-keepers, are forecast to grow to $725 M over the same period.
- HD penetration rates: there is still a lot of work to be done to develop, store, edit and transmit HD signals within both the USA and Europe.

Digital cinema is ramping up within the next five years − 10,000 US theaters are to be upgraded in 2010−2011. Digital cinemas drive significant design activity in HD and 4K video processing.

The $16 B US medical-imaging product industry will grow six percent annually in the course of 2010 based on technological advances, aging demographics and changing health care approaches. Equipment will outpace consumables, led by CT scanners and by MRI and PET machines.

All of these trends and many more lead us to believe that there is a tremendous and growing demand for a book that demystifies video processing. Professional engineers, marketers, executives and students alike need to understand:

- What is video − in terms of colors, bits and resolutions?
- What are the typical ways that video is transported?
- What functions are typical in video processing − scaling, deinterlacing, mixing?
- What are the typical challenges involved in building video-processing designs − frame buffering, line buffering, memory bandwidth, embedded control, etc.?
- What is video compression?
- How is video modulated, encoded and transmitted?

These concepts provide a solid theoretical foundation upon which the reader can build their video knowledge. This book intends to be the first text on this subject for these engineers/students.

# INTRODUCTION TO VIDEO PROCESSING

## CHAPTER OUTLINE

2.1 Digital Video: Pixels and Resolution   5
2.2 Digital Video: Pixels and Bits   6
2.3 Digital Video: Color Spaces   8
2.4 Video Processing Performance   9

Video processing — the manipulation of video to resize, clarify or compress it — is increasingly done digitally and is rapidly becoming ubiquitous in both commercial and domestic settings.

This book looks at video in the digital form — so we will talk about pixels, color spaces, etc. We start with the assumption that video is made of pixels, that a row of pixels makes a line, and a collection of lines makes a video frame. In some chapters we will briefly discuss the older analog format but mainly in the context of displaying it on a digital display.

Since this is an introductory text, and is meant to serve as a first book that clarifies digital video concepts, digital video is explained primarily through pictures, with little mathematics.

## 2.1 Digital Video: Pixels and Resolution

Digital video is made of pixels — think of a pixel as a small dot on your television screen. There are many pixels in one frame of video and many frames within one second — commonly 60 fps.

When you look at your TV there are various resolutions such as standard definition (SD), high definition (HD) with 720p or high-definition with 1080p. The resolution determines how many pixels your TV shows you. Figure 2.1 shows the number of pixels for these different resolutions — as you can see the same video frame for a 1080p TV is represented by a little over two million pixels compared to only about 300,000 pixels for standard definition. No wonder HD looks so good.

Digital Video Processing for Engineers. http://dx.doi.org/10.1016/B978-0-12-415760-6.00002-7

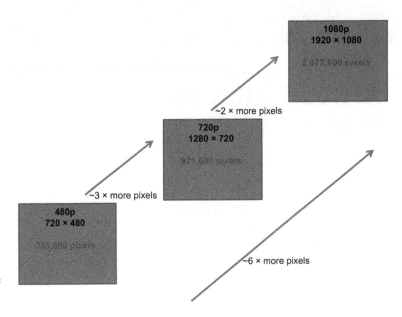

**Figure 2.1.** Increasing number of pixels in each frame of video

It might be interesting to note that the old cathode ray tube (CRT) TVs had only half of the pixels of even SD resolution shown here — so going from a CRT TV to a new 1080p TV just gave your eyes a feast of 12 times more pixels for each video frame.

The number of pixels makes a huge difference.

Take another example — when Apple created the new 'retina' display on the iPhone 4 it proved extremely popular with consumers. The new iPhone 4 had a resolution of $940 \times 640$ pixels compared to the old iPhone 3, which had a resolution of $320 \times 480$. So Apple found a way to increase the number of pixels on the same size screen by a factor of four.

The number of pixels also determines the complexity of the hardware used to manipulate these pixels. Since all manipulation is in terms of bits, let's see how pixels translate to bits.

## 2.2 Digital Video: Pixels and Bits

Each pixel has a unique color which is a combination of the primary colors: red, blue and green. How much of red, how much of blue and how much of green is the key. And this "how much" is described precisely by the value of the pixel. The value of the pixel is represented by bits and the more bits are available, the more accurate the representation. Bear in mind however, that bits are expensive to store, to manipulate and to transmit from one device to the other. So a happy balance must be realized.

Each pixel has a red (R), green (G) and blue (B) component. There are other ways to describe this as well, but we will look at red, blue and green first. Let's say that you use eight bits to store the value of red, eight bits for blue and eight bits for green. With eight bits you can have $2^8$ or 256 different possible values for red, blue and green each. When this is the case, people refer to this as a color depth of eight, or an 8-bit color depth.

Some HD video will be encoded with 10-bit color depth or even 12-bit color depth — each RGB component is encoded with 10 or 12 bits.

While more is better, remember that these bits add up. Consider 8-bit color depth. Each pixel requires $8 \times 3 = 24$ bits to represent its value.

Now think about a flat-panel TV in your house. You probably remember that this TV is 1080p — the salesperson probably also talked about $1920 \times 1080$ resolution. What this means is that each video frame shown on this flat-panel TV has 1080 lines and that each line has 1920 pixels. So you were already talking about pixels all the time — even though it may not have registered.

Let's put it together. Since each pixel requires 24 bits, and there are 1920 pixels per line and there are 1080 lines in one frame of video, this means that your hard-working flat-panel TV is showing you information that is $24 \times 1920 \times 1080 = 49{,}766{,}400$ bits in each frame. Approximately 50 million bits; also referred to as 50 Mbits. And remember most TVs go through 60 frames in one second. Some of the newer ones even go through 120 fps.

So to give you the viewing pleasure for one second we have to manipulate 3 billion bits, also referred to as 3 Gbits. And this is with 60 fps with a color depth of 8 ... It could be higher.

Table 2.1 shows the number of bits required for each frame at different resolutions. Here we have used 30 bits per pixel and also

# Table 2.1

| Image Size | Frame Size: (Total # of Pixels) | Frame Size: (Assume 30 Bits per Pixel) |
|---|---|---|
| 1920 × 1080p | 1920 × 1080 = 2 M pixels | 60 Mbits |
| 1920 × 1080i | 1920 × 1080 × 0.5 = 1 M pixels | 30 Mbits |
| 1280 × 720p | 1280 × 720 = 900 K pixels | 27 Mbits |
| SD720 × 480p | 720 × 480 × 0.5 = 173 K pixels | 5.19 Mbits |

shown the effect of interlaced video — for now just remember that the resolution is halved when the video is interlaced. The table is meant to make you aware of the amount of bits that are processed when working with digital video. Digital video processing is a demanding computational task — especially at HD resolutions. And the primary reason is the sheer number of pixels (and hence bits) involved.

## 2.3 Digital Video: Color Spaces

A color space is a method by which we can specify, create and visualize color. Each pixel has a certain color, which in simple terms can be described as a certain combination of red, blue and green. Let's represent each value of the color by eight bits. If the pixel is completely red, the R component of the pixel would be 1111 1111 and the other two components (blue and green) would be 0000 0000.

When these values are added together all we see is red. If the other two color values are not zero then the resultant color is a combination of red and some blue and some green. This color space is additive — the resultant pixel color is the sum of the intensities of each of the colors. See Figure 2.2.

The RGB color model is used to display colors on older CRT TVs as well as today's LCD TVs. Each value drives the excitations of red, green and blue phosphors on the CRT faceplate. And for digital TVs the resultant pixel value stored in hardware is converted to voltage that fires that pixel on the screen. There is more to this, including accounting for gamma correction, but we will look at that later.

Printers describe a color stimulus in terms of the reflectance and absorbance of cyan, magenta, yellow and black inks on the paper. So they work in a different color space.

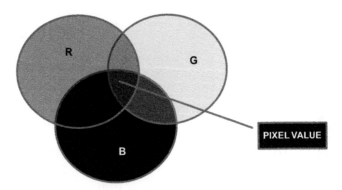

**Figure 2.2.**

$$Y = R \times 0.299 + G \times 0.587 + B \times 0.114$$
$$Cr = R \times (-0.169) + G \times (-0.332) + B \times 0.500 + 128$$
$$Cb = R \times 0.500 + G \times (-0.419) + B \times (-0.0813) + 128$$

**Figure 2.3.**

There are many color spaces — one of the more interesting ones is the YCrCb color space. This is a color space representation in which you code the pixel value in terms of its brightness (luminance), and Cr, Cb which is a combination of RGB. This method of representing color is very useful since the human eye is very sensitive to brightness or luminance, and much less sensitive to color. When the pixel value is broken down into luminance and color, we can get away with using fewer bits (lower resolution) to encode the color information as the human eye cannot detect the difference.

YCrCb is another way of encoding the RGB colors — and using fewer bits in the process — but before the video is displayed we must reconvert everything to RGB.

The way you convert a pixel value from one color space (RGB) to another (YCrCb) is to multiply each color component in the RGB space with a fixed constant — see Figure 2.3.

In terms of hardware all you need is multipliers and adders to implement the operation. Any decent processor can do this FPGAs (field-programmable gate arrays) of course can do this elegantly and very fast given their inherent DSP (digital signal processing) capabilities.

When you start converting a pixel value from one color space to another there are multiple conversions in each stage.

For example:
Convert RGB to YCrCb → TRANSMIT
→ Convert back to RGB → PROCESS THE VIDEO
→ Convert back to YCrCb → TRANSMIT
→ Convert back to RGB
→ DISPLAY

## 2.4 Video Processing Performance

Any video processing signal chain is bound to have many color space conversions along the way. These conversions have to be done at the pixel rate, which for HD video is very high.

Consider 1920 × 1080 with 60 fps. 1920 × 1080 × 60 pixels are coming in each second. Which means 124.4 million pixels in each second. In practice there is timing information associated with each frame of video which we shall ignore for now.

If this video has to be processed in real-time — which means without buffering — then the pixels have to be processed at 124.4 MHz.

Any operation that needs to be done on the bits of one pixel must be done so fast that the same operation can be done on 124.4 million pixels in the space of one second. In other words the frequency is 124.4 Mhz. In reality this is around 148 Mhz since we must account for the timing information in each video frame.

This frequency is important because whatever processing platform you choose must be able to work at this frequency. Also remember that each pixel is comprised of color planes and each color plane is represented by a certain number of bits. A color plane refers to the bits associated with each color R, G or B, for example. Let's say 8 bits for each color plane and let's assume simple RGB color planes. Going back to the processing speed, each pixel's 24 bits have to be manipulated at a frequency of 148 Mhz. With an FPGA this is relatively easy since a wide, 24-bit hardware processing chain can be laid out. If you use an 8-bit DSP, which can manipulate 8 bits, then you have to run this DSP at $3 \times 148$ Mhz to keep up with the pixels coming in. In practice HD video manipulation would normally be done on a 32-bit DSP or processor.

# SAMPLING AND ALIASING

## CHAPTER OUTLINE

3.1 Nyquist Sampling Rule   16

Whenever something is represented digitally, it must be sampled. This means that for a continuous, unbroken object to be digitized, it must be represented by many samples, close together in either time or space. If the samples are close enough, then the object appears as if it was continuous. Sampling in time, from a video perspective, means that snapshots of the scene are taken frequently enough to give an impression of continuously moving video when played back. Use of 25 to 30 images per second is common in video systems, and is generally acceptable to most viewers (this will be double in interlaced video, but that will be covered later).

Spatial sampling is applied to images, where enough pixels must be used in both vertical and horizontal dimensions to represent the image so that it appears realistic and sharp. The sampling rates can change when an image is scaled, as this is the process of converting from one resolution to another.

In order to take an analog signal and convert it to a digital signal, we need to sample the signal. For a one dimensional signal, we would use a device called an analog-to-digital converter (ADC). A camera must sample in two dimensions simultaneously, producing many samples at each instant. Unlike traditional analog cameras that expose film using chemical reactions to light, a digital camera uses a semiconductor device to convert the light into electrical signals, which can be sampled by many ADCs at the same time. A device called a charge coupled device (CCD) is commonly used. A newer alternative is to use complementary metal oxide semiconductor (CMOS) technology. Such devices can take several million samples simultaneously (this corresponds to the "Megapixel" rating used for digital cameras), and convert them into electrical signals that represent the individual pixels. These are then converted to digital representation using ADCs.

Digital Video Processing for Engineers. http://dx.doi.org/10.1016/B978-0-12-415760-6.00003-9

Color sampling is achieved by a variety of methods, but a common method is to use Bayer filters, which are color filter arrays. These can filter the incoming light at each pixel location into different colors — such as red, green and blue — and record the pixel intensities of each color separately. Newer methods build color sensitive sensors into the CMOS technology. There are many other factors affecting this process, such as aperture, shutter speed, zooming and focusing, which are not discussed here.

To explain some of the concepts of sampling and aliasing, we will consider a one-dimensional signal for simplicity. The ADC will measure the signal at rapid intervals, and produce samples (pixels in the image and video domain). It will output a digital signal proportional to the amplitude of the analog signal at that instant. This can be compared to looking at an object with only a strobe light for illumination. You can see the object only when the strobe light flashes. If the object is not moving, then everything looks pretty much the same as if we used a normal, continuous light source. Where things get interesting is when we look at a moving object with the strobe light. If the object is moving rapidly, then the appearance of the motion can be quite different from that when viewed under normal light. We can also see strange effects even if the object is moving fairly slowly, or we reduce the rate of the strobe light enough. Intuitively, we can see that what is important is the rate of the strobe light compared to the rate of movement of the illuminated object. As long as the light strobes fast enough compared to the movement of the object, this movement looks very fluid and normal. When the light strobes slowly compared to the rate of object movement, the movement looks odd, often like slow motion, as we can see the object is moving, but we miss the sense of continuous, fluid movement.

Let's mention one more example: a simple animated movie such as sketching a character on index cards. To depict this character moving, perhaps jumping and falling, we might sketch 20 or 40 cards, each showing the same character in sequential stages of this motion, with just small movement changes from one card to next. Then, by holding one edge of the deck of index cards and flipping through it quickly by thumbing the other edge, the character appears to continuously jump and fall. This is the process used for video — the screen is being updated at about 30 times per second, which is rapid enough for us not to notice the separate frames, so the motion appears to be continuous.

So it makes sense that if we sample a signal very fast compared to how rapidly the signal is changing, we get a pretty accurate sampled representation of the signal, but if we sample too slow, we will see a distorted version of the signal.

Below are graphs of two different sinusoidal signals being sampled. The slower-moving signal (lower frequency) in Figure 3.1 can be represented accurately with the indicated sample rate, but the faster-moving signal (higher frequency) in Figure 3.2 is not accurately represented by our sample rate. In fact, it actually appears to be a slow-moving (low frequency) signal, as indicated by the dashed line. This shows the importance of sampling fast enough for a given frequency signal.

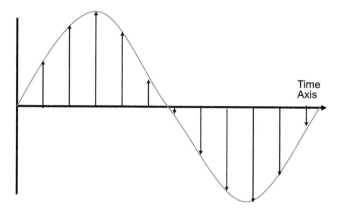

Time
Axis

**Figure 3.1.** Sampling a low-frequency signal (arrows indicate sample instants).

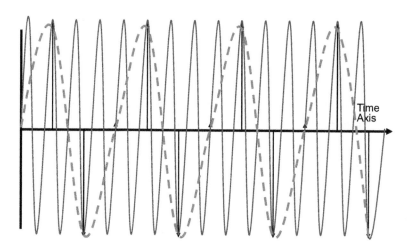

Time
Axis

**Figure 3.2.** Sampling a high-frequency signal (same sample instants).

The dashed line shows how the sampled signal will appear if we connect the sample dots and smooth out the signal. Notice that since the actual (solid line) signal is changing so rapidly between sampling instants, this movement is not apparent in the sampled version of the signal. The sampled version appears to be a lower-frequency signal than the actual signal. This effect is known as **aliasing**.

We need a way to quantify how fast we must sample to accurately represent a given signal. We also need to better understand exactly what is happening when aliasing occurs. It may seem strange, but there are some instances when aliasing can be useful.

Let's go back to the analogy of the strobe light, and try another thought experiment. Imagine a spinning wheel, with a single dot near the edge. Let's set the strobe light to flash every $^1/_8$ of a second, or eight times per second. Below is shown what we see over six flashes, depending on how fast the wheel is rotating. **Time increments from left to right in all figures.**

The reality is that once we sample a signal (this is what we are doing by flashing the strobe light), we cannot be sure what has happened in-between flashes. Our natural instinct is to assume that the signal (or dot in our example) took the shortest path from where it appears in one flash to its position in the subsequent flash. But, as we can see in the examples above, this can be misleading. The dot could be moving around the circle in the opposite direction (taking the longer path) to get to the point where we see it on the next flash. Or imagine that the circle is rotating in the assumed direction, but it rotates one full revolution plus "a little bit extra" every flash (the 9 Hz diagram). What

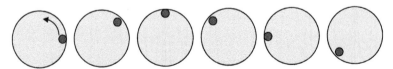

Wheel rotating counterclockwise once per second, or 1 Hertz (Hz)

**Figure 3.3.** The dot moves $^1/_8$ of a revolution with each strobe flash.

**Figure 3.4.** Now the dot moves twice as fast, ¼ of a revolution with each strobe flash.

Wheel rotating counterclockwise twice per second, or 2 Hz

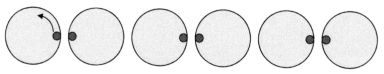

Wheel rotating counterclockwise four times per second, or 4 Hz

**Figure 3.5.** The dot moves ½ of a revolution with each strobe flash. It appears to alternate on each side of the circle with each flash. Can you be sure which direction the wheel is rotating?

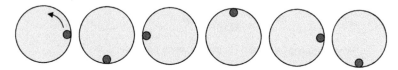

Wheel rotating counterclockwise six times per second, or 6 Hz

**Figure 3.6.** The dot moves counterclockwise ¾ of a revolution with each strobe flash. But it appears to be moving backwards (clockwise).

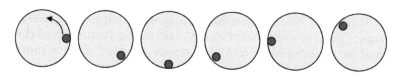

Wheel rotating counterclockwise seven times per second, or 7 Hz

**Figure 3.7.** The dot moves almost a complete revolution counter-clockwise, ⅞ revolution with each strobe flash. Now it definitely appears to be moving backwards (clockwise).

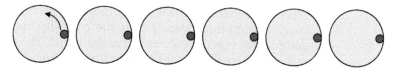

Wheel rotating counterclockwise eight times per second, or 8 Hz

**Figure 3.8.** It looks like the dot has stopped moving! What is happening is that the dot completes exactly one revolution every strobe interval. You can't tell whether the wheel is moving at all, spinning forwards or backwards. In fact, could the wheel be rotating twice per strobe interval?

we see is only the "a little bit extra" on every flash. For that matter, it could go around 10 times plus the same "a little bit extra" and we couldn't tell the difference.

**Figure 3.9.** It sure looks the same as when the wheel was rotating once per second, or 1 Hz. In fact, the wheel is moving $1\frac{1}{8}$ revolution with each strobe flash.

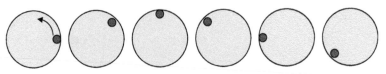

Wheel rotating counterclockwise nine times per second, or 9 Hz

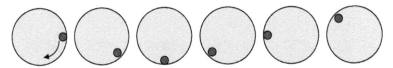

Wheel rotating backward (clockwise) once per second, or -1 Hz

**Figure 3.10.** Now we stopped the wheel and started rotating backward at $-\frac{1}{8}$ revolution with each strobe flash. Notice that this appears exactly the same as when we were rotating forward at $\frac{7}{8}$ revolution with each strobe flash.

## 3.1 Nyquist Sampling Rule

To prevent all this, we have to come up with a sampling rule, or convention. What we are going to agree is that we will always sample (or strobe) at least twice as fast as the frequency of the signal we are interested in. And in reality, we need to have some margin, so we better make sure we are sampling *more* than twice as fast as our signal.

Consider what happens when we start the wheel moving slowly in a counterclockwise direction. Everything looks fine until we reach a rotational speed of 4 Hz. At this point the dot will appear to be alternating on either side of the circle with each strobe flash. Once we have reached this point, we can no longer tell which direction the wheel is rotating — it will look the same rotating both directions. This is the critical point, where we are sampling at exactly twice as fast as the signal. The sampling speed is the frequency of the strobe light (this would be analogous to the ADC sample frequency), eight times per second, or 8 Hz. The rotational speed of the wheel (our signal) is 4 Hz.

If we spin the wheel any faster, it appears to move backwards (clockwise), and by the time we reach a rotational speed of 8 Hz, it appears to stop altogether. Spinning still faster will make the wheel appear to move forwards again, until it again seems to move backwards and the cycle repeats.

To summarize, whenever you have a sampled signal, you cannot be completely sure of its frequency. But if you assume that the rule was followed — that the signal was sampled at more than

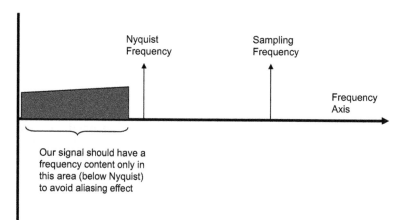

**Figure 3.11.** Nyquist Sampling Rule.

twice the frequency of the signal — then the sampled signal will represent the same frequency as the actual signal prior to sampling. The critical frequency that the signal must never exceed, which is one half of the sampling frequency, is called the *Nyquist* frequency (see Figure 3.11).

If we follow this rule, then we can avoid the aliasing phenomenon we demonstrated with the moving wheel example above. Normally, the ADC converter frequency is set high enough to sample the signal upon which we want to perform digital signal processing. To make sure that unwanted signals above the Nyquist frequency do not get sampled and cause aliasing there are two options:

The analog signal needs to be filtered through an analog low-pass filter, which attenuates any unwanted high-frequency content signals, just prior to the sampling.

The sampling speed is set at least twice as fast as the fastest-changing components in the analog signal.

A common one-dimensional example is the telephone system. Our voices are assumed to have a maximum frequency of about 3600 Hz. At the microphone, our voice is filtered by an analog filter to eliminate, or at least substantially reduce, any frequencies above 3600 Hz. Then the microphone signal is sampled at 8000 Hz, or 8 kHz. All subsequent digital signal processing occurs on this 8 kSPS (kilo-samples per second) signal. That is why if you hear music in the background while on the telephone, the music will sound flat or distorted. Our ears can detect up to about 15 kHz frequency, and music generally has frequency content exceeding 3600 Hz. But little of this higher frequency content will pass through the telephone system.

Images are sampled spatially, in vertical and horizontal directions. In video sampling, we add the time, or temporal dimension, where the sampling rate is equal to the frame rate. Intensity and hues (YCrCb) or colors (RGB) are also sampled separately and sometimes at different resolutions.

There remains one last issue related to sampling, which is quantization, which is concerned with how many bits are used to represent each sample. For video processing we need to have a sufficient number of bits to represent the color and intensity differences that our eyes are capable of discerning. The most common is to use 10 bits per sample, which results in $2^{10}$ or 1024 different levels per sample. Other common sample resolutions are 8- and 12-bits, giving 256 or 4096 levels respectively. These 10−12 bit resolutions are normally sufficient for video. Where quantization becomes important is in video compression, which will be covered in subsequent chapters.

# INTRODUCTION TO DIGITAL FILTERING

## CHAPTER OUTLINE

4.1 Median Filtering  19
4.2 FIR Filters  20
4.3 FIR Filter Construction  20
4.4 Computing Frequency Response  23

Digital filtering is used in different aspects of video signal processing. One application is reducing random noise effects or corrupted pixels in images or video. Finite Impulse Response (FIR) filters can be used to smooth out an image, and to reduce random noise, but the drawback is that it will soften or blur sharp edges in the image.

## 4.1 Median Filtering

Another type of filter often better suited to reducing noise in an image is the median filter. A median filter is a non-linear filter, which does not use multipliers. It analyzes the image pixel by pixel, and replaces each pixel with the median of neighboring entries. The pattern of neighboring pixels represents a window, which slides, entry by entry, over the entire image. The window is usually a box or cross pattern, centered on the pixel being analyzed. Since the window has an odd number of entries, then the median is simple to define: it is just the middle value after all the entries in the window are sorted numerically.

Median filtering is used as a smoothing technique, which is effective at removing noise in smooth patches or smooth regions of a signal. Unlike low-pass FIR filters, the median filter tends to preserve the edges in an image. Because of this, median filtering is very widely used in digital image processing.

Digital Video Processing for Engineers. http://dx.doi.org/10.1016/B978-0-12-415760-6.00004-0

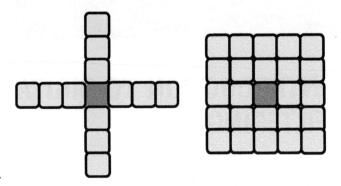

**Figure 4.1.** Examples of median filter patterns.

## 4.2 FIR Filters

Scaling, or changing image resolution, is an application which requires linear filtering. When upsampling, or increasing resolution, a FIR filter is normally used to perform the interpolation. When downsampling, the image is often low-pass filtered first to remove the higher spatial frequencies that may alias with reduced resolution.

## 4.3 FIR Filter Construction

Let's begin with how to construct a FIR filter. A FIR filter is built of multipliers and adders, can be implemented in hardware or software, and run in a serial fashion, parallel fashion, or some combination. We will focus on the parallel implementation, because it's the most straightforward to understand.

A key property of an FIR filter is the number of taps or multipliers required to compute each output. In a parallel implementation, the number of taps equals the number of multipliers. In a serial implementation, one multiplier is used to perform the multiple operations sequentially for each output. Assuming single clock-cycle multipliers, a parallel FIR filter can produce one output each clock-cycle, and a serial FIR filter would require N clock-cycles to produce each output, where N is the number of filter taps. Figure 4.2 shows a small 5-tap parallel filter.

The inputs and outputs of the FIR filter are sampled data. For simplicity, we will assume that the inputs, outputs and filter coefficients $C_m$ are all real numbers. The input data stream will be denoted as $x_k$, and the output $y_k$. The "k" subscript is used to identify the sequence of data. For example, $x_{k+1}$ follows $x_k$, and $x_{k-1}$ precedes $x_k$. For the purpose of defining a steady state response, we often assume that the data streams are infinitely long in time, or that k extends from $-\infty$ to $+\infty$.

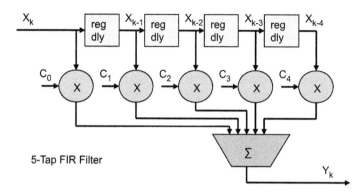

**Figure 4.2.** FIR filter diagram.

The coefficients are usually static (they do not change over time), and determine the filter's frequency response.

In equation form, the filter could be represented as:

$$y_k = C_0 \times x_k + C_1 \times x_{k-1} + C_2 \times x_{k-2} + C_3 \times x_{k-3} + C_4 \times x_{k-4}$$

This is just the sum of the multipliers. This could get rather tedious to write as the number of taps gets larger, so the following short-hand summation is often used:

$$y_k = \sum_{i=0 \text{ to } 4} C_i \, x_{k-i}$$

We can also make the equation for any length of filter. To make our filter of length "N", we simply replace the 4 (5-1 taps) with $N - 1$.

$$y_k = \sum_{i=0 \text{ to } N-1} C_i \, x_{k-i}$$

Another way to look at this is that the data stream $x_{k+2}$, $x_{k+1}$, $x_k$, $x_{k-1}$, $x_{k-2}$... is sliding past a fixed array of coefficients. At each clock-cycle, the data and coefficients are cross-multiplied and the outputs of all multipliers for that clock-cycle are summed to form a single output (this process also known as dot product). Then, on the next clock-cycle, the data is shifted one place relative to the coefficients (which are fixed), and the process repeated. This process is known as convolution.

The FIR structure is very simple, yet has the ability to create almost any frequency response, given a sufficient number of taps. This is very powerful, but unfortunately not at all intuitive. It's somewhat analogous to the brain — a very simple structure of interconnected neurons, yet the combination can produce amazing results. In the course of the rest of this chapter, we will try to gain some understanding of how this happens.

Below is an example using actual numbers, to illustrate this process called convolution.

We will define a filter of 5 coefficients $\{C_0, C_1, C_2, C_3, C_4\} = \{1,3,5,3,1\}$.

Our $x_k$ sequence will be defined as $\{x_0, x_1, x_2, x_3, x_4, x_5\} = \{-1, 1, 2, 1, 4, -1\}$ and $x_k = 0$ for $k < 0$ and for $k > 6$ (everywhere else).

Let's start by computing $y_{-1} = \sum_{i=0 \text{ to } N-1} C_i\, x_{-1-i}$

We can see that the subscript on x will be negative for all $i = 0$ to 4. In this example $y_k = 0$ for $k < 0$. That is, until there is a non-zero input $x_k$, the output $y_k$ will also be zero. Things start to happen at $k = 0$, because $x_0$ is the first non-zero input.

$$y_{-1} = \sum_{i=0 \text{ to } N-1} C_i\, x_{-1-i} = (1)(0) + (3)(0) + (5)(0) + (3)(0) + (1)(0) = 0$$

$$y_0 = \sum_{i=0 \text{ to } N-1} C_i\, x_{0-i} = (1)(-1) + (3)(0) + (5)(0) + (3)(0) + (1)(0) = -1$$

$$y_1 = \sum_{i=0 \text{ to } N-1} C_i\, x_{1-i} = (1)(1) + (3)(-1) + (5)(0) + (3)(0) + (1)(0) = -2$$

$$y_2 = \sum_{i=0 \text{ to } N-1} C_i\, x_{2-i} = (1)(2) + (3)(1) + (5)(-1) + (3)(0) + (1)(0) = 0$$

$$y_3 = \sum_{i=0 \text{ to } N-1} C_i\, x_{3-i} = (1)(1) + (3)(2) + (5)(1) + (3)(-1) + (1)(0) = 9$$

$$y_4 = \sum_{i=0 \text{ to } N-1} C_i\, x_{4-i} = (1)(\mathbf{4}) + (3)(1) + (5)(2) + (3)(1) + (1)(-1) = 19$$

$$y_5 = \sum_{i=0 \text{ to } N-1} C_i\, x_{5-i} = (1)(-1) + (3)(\mathbf{4}) + (5)(1) + (3)(2) + (1)(1) = 23$$

$$y_6 = \sum_{i=0 \text{ to } N-1} C_i\, x_{6-i} = (1)(0) + (3)(-1) + (5)(\mathbf{4}) + (3)(1) + (1)(2) = 22$$

$$y_7 = \sum_{i=0 \text{ to } N-1} C_i\, x_{7-i} = (1)(0) + (3)(0) + (5)(-1) + (3)(\mathbf{4}) + (1)(1) = 8$$

$$y_8 = \sum_{i=0 \text{ to } N-1} C_i\, x_{8-i} = (1)(0) + (3)(0) + (5)(0) + (3)(-1) + (1)(\mathbf{4}) = 1$$

$$y_9 = \sum_{i=0 \text{ to } N-1} C_i\, x_{9-i} = (1)(0) + (3)(0) + (5)(0) + (3)(0) + (1)(-1) = -1$$

$$y_{10} = \sum_{i=0 \text{ to } N-1} C_i\, x_{10-i} = (1)(0) + (3)(0) + (5)(0) + (3)(0) + (1)(0) = 0$$

$$y_{11} = \sum_{i=0 \text{ to } N-1} C_i\, x_{11-i} = (1)(0) + (3)(0) + (5)(0) + (3)(0) + (1)(0) = 0$$

$$y_{12} = \sum_{i=0 \text{ to } N-1} C_i\, x_{12-i} = (1)(0) + (3)(0) + (5)(0) + (3)(0) + (1)(0) = 0$$

This is definitely tedious. There are a couple of things to notice. Follow the input $x_4 = 4$ (in bold) in our example. See how it moves across, from one multiplier to the next. Each input sample $x_k$ will be multiplied by each tap in turn. Once it passes through the filter, that input sample is discarded and has no further influence upon the output. In our example, $x_4$ is discarded after computing $y_8$.

Once the last non-zero input data $x_k$ has shifted all the way through the filter taps, the output data $y_k$ will go to zero (this starts at $k = 10$ in our example).

Now let's consider a special case, where $x_k = 1$ for $k = 0$, and $x_k = 0$ for $k \neq 0$. This means that we only have one non-zero input sample, and it is equal to 1. Now if we compute the output, which is simpler this time, we get:

$$y_{-1} = \sum_{i=0 \text{ to } N-1} C_i \; x_{-1-i} = (1)(0) + (3)(0) + (5)(0) + (3)(0) + (1)(0) = 0$$

$$y_0 = \sum_{i=0 \text{ to } N-1} C_i \; x_{0-i} = (1)(1) + (3)(0) + (5)(0) + (3)(0) + (1)(0) = 1$$

$$y_1 = \sum_{i=0 \text{ to } N-1} C_i \; x_{1-i} = (1)(0) + (3)(1) + (5)(0) + (3)(0) + (1)(0) = 3$$

$$y_2 = \sum_{i=0 \text{ to } N-1} C_i \; x_{2-i} = (1)(0) + (3)(0) + (5)(1) + (3)(0) + (1)(0) = 5$$

$$y_3 = \sum_{i=0 \text{ to } N-1} C_i \; x_{3-i} = (1)(0) + (3)(0) + (5)(0) + (3)(1) + (1)(0) = 3$$

$$y_4 = \sum_{i=0 \text{ to } N-1} C_i \; x_{4-i} = (1)(0) + (3)(0) + (5)(0) + (3)(0) + (1)(1) = 1$$

$$y_5 = \sum_{i=0 \text{ to } N-1} C_i \; x_{5-i} = (1)(0) + (3)(0) + (5)(0) + (3)(0) + (1)(0) = 0$$

$$y_6 = \sum_{i=0 \text{ to } N-1} C_i \; x_{6-i} = (1)(0) + (3)(0) + (5)(0) + (3)(0) + (1)(0) = 0$$

Notice that the output is the same sequence as the coefficients. This should come as no surprise once you think about it. This output is defined as the filter's impulse response, so named as it occurs when the filter input is an impulse, or a single non-zero input equal to one. This gives the FIR filter its name. By Finite Impulse Response" or FIR, this indicates that if this type of filter is driven with an impulse, we will see a response (the output) has a finite length, after which it becomes zero. This may seem trivial, but it is a very good property to have, as we will see in the chapter on infinite impulse response filters.

## 4.4 Computing Frequency Response

So far we have covered the mechanics of building the filter, and how to compute the output data, given the coefficients and input data. But we do not have any intuitive feeling as to how this operation can allow some frequencies to pass through, and block other frequencies. A very basic understanding of a low pass filter can be gained by the concept of averaging. We all know that if we average multiplication results, we get a smoother, more consistent output, as rapid fluctuations are damped out. A moving average filter is simply a filter with all the coefficients set to 1. The more filter taps, the longer the averaging, and the more smoothing takes place. This gives an idea of how a filter structure can remove high frequencies, or rapid fluctuations. Now imagine if the filter taps were alternating $+1$, $-1$, $+1$, $-1$ ... and so on. A slowly varying input signal will have adjacent samples nearly the same, and these will cancel in the filter, resulting in a nearly zero output. This filter is blocking low frequencies. On the other hand, an input signal near the Nyquist rate will have big changes from sample to sample, and will result in a much larger output. To get a more precise handle on how to configure the coefficient values to get the desired frequency response however, we need to use a bit of math.

We will start by computing the frequency response of the filter from the coefficients. Remember, the frequency response of the filter is determined by the coefficients (also called the impulse response).

Let's begin by trying to determine the frequency response of a filter by measurement. Imagine if we take a complex exponential signal of a given frequency, and use this as the input to our filter. Then we measure the output. If the frequency of the exponential signal is in the passband of the filter, it will appear at the output. But if the frequency of the exponential signal is in the stopband of the filter, it will appear at the output with a much lower level than the input, or not at all. Imagine we start with a very low frequency exponential input, and take this measurement, then slightly increase the frequency of the exponential input, measure again, and keep going until the exponential frequency is equal to the Nyquist frequency. If we plot the level of the output signal across the frequency from 0 to $F_{Nyquist}$, we will have the frequency response of the filter. It turns out that we don't have to take all these measurements as we can compute this fairly easily as shown below.

Let's review the last equation. It's just a sampled version of a signal rotating around the unit circle. We sample at time $= m$,

| | |
|---|---|
| $y_k = \sum_{i=0 \text{ to } 4} c_i\, x_{k-i}$ | Output of our five-tap example filter. |
| $y_k = \sum_{i=-\infty \text{ to } \infty} c_i\, x_{k-i}$ | Same equation, except that we are allowing an infinite number of coefficients (no limits on filter length). |
| $x_m = e^{j\omega m} = \cos(\omega m) + j\, \sin(\omega m)$ | This is our complex exponential input at $\omega$ radians per sample. |

and then sample again at time $= m+1$. So from one sample to the next, our sampled signal will move $\omega$ radians around the unit circle. If we are sampling at 10 times faster than we are moving around the unit circle, then it will take 10 samples to get around the circle, and move $2\pi / 10$ radians each sample.

$x_m = e^{j2\pi m / 10} = \cos(2\pi m / 10) + j \sin(2\pi m / 10)$
when $\omega = 2\pi / 10$

To clarify, Table 4.1 shows $x_m = e^{j2\pi m/10}$ evaluated at various "m" using a complex exponential signal, which is a rotating vector in the frequency domain. If you want to check using a calculator, remember that the angles are in units of radians, not degrees.

# Table 4.1

| | | |
|---|---|---|
| $m = 0$ | $x_0 = e^{j0} = \cos(0) + j \sin(0)$ | $1 + j0$ |
| $m = 1$ | $x_1 = e^{j\pi/5} = \cos(\pi/5) + j \sin(\pi/5)$ | $0.8090 + j0.5878$ |
| $m = 2$ | $x_2 = e^{j2\pi/5} = \cos(2\pi/5) + j \sin(2\pi/5)$ | $0.3090 + j0.9511$ |
| $m = 3$ | $x_3 = e^{j3\pi/5} = \cos(3\pi/5) + j \sin(3\pi/5)$ | $-0.3090 + j0.9511$ |
| $m = 4$ | $x_4 = e^{j4\pi/5} = \cos(4\pi/5) + j \sin(4\pi/5)$ | $-0.8090 + j0.5878$ |
| $m = 5$ | $x_5 = e^{j\pi} = \cos(\pi) + j \sin(\pi)$ | $-1 + j0$ |
| $m = 6$ | $x_6 = e^{j6\pi/5} = \cos(6\pi/5) + j \sin(6\pi/5)$ | $-0.8090 - j0.5878$ |
| $m = 7$ | $x_7 = e^{j7\pi/5} = \cos(7\pi/5) + j \sin(7\pi/5)$ | $-0.3090 - j0.9511$ |
| $m = 8$ | $x_8 = e^{j8\pi/5} = \cos(8\pi/5) + j \sin(8\pi/5)$ | $0.3090 - j0.9511$ |
| $m = 9$ | $x_9 = e^{j9\pi/5} = \cos(9\pi/5) + j \sin(9\pi/5)$ | $0.8090 - j0.5878$ |
| $m = 10$ | $x_{10} = x_0 = e^{j2\pi} = \cos(2\pi) + j \sin(2\pi)$ | $1 + j0$ |

We could next increase $x_m$ so that we rotate the unit circle every five samples. This is twice as fast as before.

$x_m = e^{j2\pi m / 5} = \cos(2\pi m / 5) + j \sin(2\pi m / 5)$ when $\omega = 2\pi / 5$

Now we go back to the filter equation, and substitute the complex exponential input for $x_{k-i}$.

$y_k = \sum_{i = -\infty \text{ to } \infty} C_i x_{k-i}$
$x_m = e^{j\omega m} = \cos(\omega m) + j \sin(\omega m)$

Insert k-i for m

$x_{k-i} = e^{j\omega(k-i)} = \cos(\omega(k-i)) + j\sin(\omega(k-i))$

Next, replace in $x_{k-i}$ in filter equation:

$y_k = \sum_{i = -\infty \text{ to } \infty} C_i e^{j\omega(k-i)}$

There is a property of exponentials that we frequently need to use:

$e^{(a+b)} = e^a \times e^b$ and $e^{(a-b)} = e^a \times e^{-b}$

If you remember your scientific notation, this makes sense. For example:

$$10^2 \times 10^3 = 100 \times 1000 = 100{,}000 = 10^5 = 10^{(2+3)}$$

Now back to the filter equation:

$$y_k = \sum_{i=-\infty \text{ to } \infty} C_i \, e^{j\omega(k-i)} = \sum_{i=-\infty \text{ to } \infty} C_i \, e^{j\omega k} \times e^{-j\omega i}$$

Let's do a little algebra trick. Notice the term $e^{j\omega k}$ does not contain the term i used in the summation. So we can pull this term out in front of the summation.

$$y_k = e^{j\omega k} \times \sum_{i=-\infty \text{ to } \infty} C_i \, e^{-j\omega i}$$

Notice the term $e^{j\omega k}$ is just the complex exponential we used as an input.

$$y_k = x_k \times \sum_{i=-\infty \text{ to } \infty} C_i \, e^{-j\omega i}$$

Voila! The expression $\sum_{i=-\infty \text{ to } \infty} C_i \, e^{-j\omega i}$ gives us the value of the frequency response of the filter at frequency $\omega$. It is solely a function of $\omega$ and the filter coefficients.

This expression applies a gain factor to the input, $x_k$, to produce the filter output. Where this expression is large, we are in the passband of the filter. If this expression is close to zero, we are in the stopband of the filter.

Let's give this expression a less cumbersome representation. Again, it is a function of $\omega$, which we expect, because the characteristics of the filter vary with frequency. It is also a function of the coefficients, $C_i$, but these are assumed to be fixed for a given filter.

**Frequency response = $H(\omega) = \sum_{i=-\infty \text{ to } \infty} C_i \, e^{-j\omega i}$**

Now in reality, it is not as bad as it looks. This is the generic version of the equation, where we must allow for an infinite number of coefficients (or taps). But suppose we are determining the frequency response of our 5 tap example filter.

$$H(\omega) = \sum_{i=0 \text{ to } 4} C_i \, e^{-j\omega i} \text{ and } \{C_0, C_1, C_2, C_3, C_4\} = \{1, 3, 5, 3, 1\}$$

Let's find the response of the filter at a couple of different frequencies. First, let $\omega = 0$. This corresponds to DC input — we are putting a constant level signal into the filter. This would be $x_{k=1}$ for all values k.

$$H(0) = C_0 + C_1 + C_2 + C_3 + C_4 = 1 + 3 + 5 + 3 + 1 = 13$$

This one was simple, since $e^0 = 1$. The DC or zero frequency response of the filter is called the gain of the filter. Often, it may be convenient to force the gain $= 1$, which would involve dividing all the individual filter coefficients by $H(0)$. The passband and stopband characteristics are not altered by this process, since all the coefficients are scaled equally. It just normalizes the frequency response so the passband has a gain equal to one.

Now we will compute the frequency response for $\omega = \pi / 2$.

$$H(\pi/2) = C_0\,e^0 + C_1\,e^{-\pi/2} + C_2\,e^{-\pi} + C_3\,e^{-3\pi/2} + C_4\,e^{-4\pi/2}$$
$$= 1 \times 1 + 3 \times (-j) + 5 \times (-1) + 3 \times (j) + 1 \times 1 = -3$$

So the magnitude of the frequency response has gone from 13 (at $\omega = 0$) to 3 (at $\omega = \pi / 2$). The phase has gone from 0 degrees (at $\omega = 0$) to 180 degrees (at $\omega = \pi / 2$), although generally we are not concerned about the phase response of FIR filters. Just from these two points of the frequency response, we can guess that the filter is probably some type of low pass filter.

The magnitude is calculated as follows:

Magnitude $Z = X + jY = (X^2 + Y^2)^{1/2} = |\,Z\,|$

Our example calculation above turned out to have only real numbers, but that is because the imaginary components of $H(\pi / 2)$ canceled out to zero. The magnitude of $H(\pi / 2)$ is:

Magnitude $|\text{-}3 + 0j| = 3$

A computer program can easily evaluate $H(\omega)$ from $-\pi$ to $\pi$ and plot it for us. Of course, this is almost never done by hand. Figure 4.3 is a frequency plot of this filter using a FIR Filter program.

Not the best filter, but it is still a low-pass filter. The frequency axis is normalized to $F_s$, and the magnitude of the amplitude is

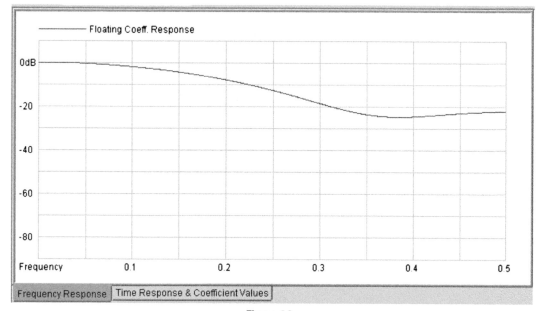

**Figure 4.3.**

plotted on a logarithmic scale, referenced to a passband frequency response of one.

We can verify our hand calculation was correct. We calculated the magnitude of $|H(\pi/2)|$ at 3, and $|H(0)|$ at 13. The logarithmic difference is:

$20 \log_{10}(3/13) = -12.7$ dB

If you check the frequency response plot above, you will see at frequency $F_s/4$ (or 0.25 on the normalized frequency scale), which corresponds to $\pi/2$ in radians, the filter does indeed seem to attenuate the input signal by about 12 to 13 dB relative to $|H(0)|$. Other filter programs might plot the frequency axis referenced from 0 to $\pi$, or from $-\pi$ to $\pi$.

# 5

# VIDEO SCALING

## CHAPTER OUTLINE

5.1 Understanding Video Scaling   30
5.2 Implementing Video Scaling   33
5.3 Video Scaling for Different Aspect Ratios   36
5.4 Conclusion   38

Video comes in different sizes — as anyone who has watched standard definition DVDs on their high resolution HDTVs has no doubt experienced. Each video frame, which consists of lines of pixels, has a different size as shown in Figure 5.1.

Video has to be resized to view it on different sized displays — this means that the native resolution of the video frame is adjusted to fit the display available. Video resizing or scaling is an increasingly common function used to convert images of one resolution and aspect ratio to another "target" resolution and/or aspect ratio. The most familiar example of video scaling is scaling a VGA signal (640 × 480) output from a standard laptop to an SXGA signal (1280 × 1024) for display on LCD monitors.

For high-volume systems dealing with standardized image sizes such as HD television, video scaling is most efficiently done using application specific standard products (ASSPs). However, many video applications such as video surveillance, broadcast display and monitoring, video conferencing and specialty displays, need solutions that can handle custom image sizes and differing levels of quality. This often requires custom scaling algorithms. FPGAs with an array of high-performance DSP structures are ideally suited for such algorithms, and FPGA vendors are beginning to offer user-customizable video-scaling IP blocks that can be quickly configured for any application.

Scaling is often combined with other algorithms such as deinterlacing and aspect ratio conversion. In this chapter we will focus on the digital video processing involved in resizing or scaling a video frame.

Towards the end of this chapter we will briefly describe the different aspect ratios since scaling is done many times to stretch a 4:3 image to display on today's 16:9 HDTV displays.

Digital Video Processing for Engineers. http://dx.doi.org/10.1016/B978-0-12-415760-6.00005-2

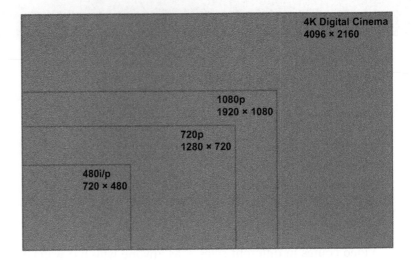

**Figure 5.1.**

## 5.1 Understanding Video Scaling

Video scaling, whether upscaling or downscaling, is the process of generating pixels that did not exist in the original image. To illustrate this, let's look at a simple example: scaling a 2 × 2 pixel image to a 4 × 4 pixel image, as shown in Figure 5.2. In the "New Image" the white pixels are pixels from the "Existing Image" and the black pixels are those that need to be generated from the existing pixels.

There are many methods for generating new pixels: the simplest is called the nearest neighbor method, or 1 × 1 interpolation. In this method a new pixel value is simply equal to the

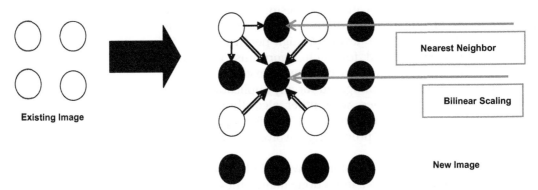

Existing Image

Nearest Neighbor

Bilinear Scaling

New Image

**Figure 5.2.** A 2 x 2 pixel image is enlarged (upscale) to a 4 x 4 pixel image.

value of the preceding pixel. This is the simplest example of scaling and requires minimal hardware resources since no calculations have to be done

In a slightly more sophisticated approach, called bilinear scaling, a new pixel is equal to the average of the two neighboring pixels in both the vertical and horizontal dimensions. Both of these techniques are illustrated conceptually in Figure 5.2.

Bilinear scaling implicitly assumes equal weighting of the four neighboring pixels, i.e. each pixel is multiplied by 0.25.

New Pixel $= (0.25 \times$ pixel 1$) + (0.25 \times$ pixel 1$) + (0.25 \times$ pixel 1$) + (0.25 \times$ pixel 1$)$

By creating new values for each new pixel in this way the image size is doubled – a frame size of $2 \times 2$ is increased to a frame size of $4 \times 4$. The same concept can be applied to larger frames.

In video terminology this function is scalar (all coefficients are equal) – it has two taps in the horizontal dimension, two taps in the vertical dimension and has one phase (i.e. one set of coefficients).

The term "taps" refers to filter taps, as scaling is mathematically identical to generalized filtering, i.e. multiplying coefficients by inputs (taps) and summing, such as a direct form digital filter. The hardware structure that implements this bilinear scaling function is simplistically shown in Figure 5.3. In practice it can be optimized to utilize less hardware resources.

To further illustrate this, consider the $4 \times 4$ scaler shown in Figure 5.4. Four new (black) pixels are generated for every one existing (white) pixel. Like the previous example, it has two taps in the horizontal dimension. Unlike the previous example, each new pixel is generated using a different weighting of the two existing pixels, making this a four-phase scaler.

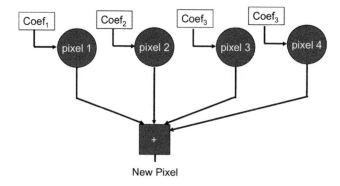

Figure 5.3.

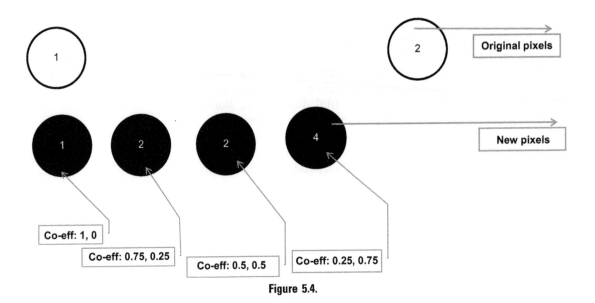

**Figure 5.4.**

The weighting coefficients for each pixel are given below.

New pixel 1 will just be a copy of the original pixel 1.

New pixel 2 will weigh 75% of original pixel 1 and 25% of original pixel 2.

New pixel 3 will weigh 50% of original pixel 1 and 50% of original pixel 2.

New pixel 4 will weigh 25% of original pixel 1 and 75% of original pixel 2.

These are the simplest scaling algorithms and in many cases these suffice for video scaling.

It's easy to see how this method of creating a pixel can be increased in complexity by using more and more original pixels. For example, you can choose to use nine original pixels each multiplied by a different coefficient to calculate the value of your new pixel.

Figure 5.5 shows an example of what is known as bicubic scaling, or $4 \times 4$ scaling. A sample $4 \times 4$ matrix of image pixels is downscaled by a factor of four in both the horizontal and the vertical dimension — so a $4 \times 4$ matrix of pixels is reduced to a single pixel. The four pixels in the vertical dimension are scaled first — these four pixels belong to four different lines of video in that frame.

By using any appropriate set of coefficients, the four pixel values are converted into a single pixel value. This is then repeated for the next four pixels, and so on. Finally we are left with four pixels. These are further reduced to a single pixel by

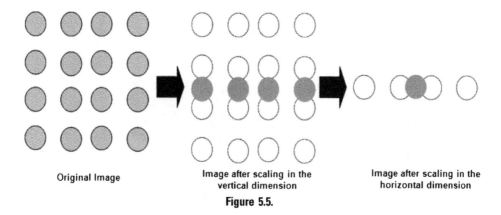

Original Image | Image after scaling in the vertical dimension | Image after scaling in the horizontal dimension

**Figure 5.5.**

applying yet another set of coefficients to these pixels — yielding a 16:1 downscaling for this video stream.

## 5.2 Implementing Video Scaling

You will recall that video scaling is mathematically equivalent to digital filtering since we are multiplying a pixel value by a coefficient and then summing up all the product terms. The implementation is thus very similar to the implementation of two 1-D filters.

Let's stay with the example we used in Figure 5.5.

First we will need to store the four lines of video. While we are working on only four pixels — one from each line — in practice the entire video line will have to be stored on-chip. This will account for an appreciable amount of memory. For example, in a 1080p video frame, each video line means 1920 pixels with, for example, each pixel requiring 24 bits = 46 Kbits or 5.7 KB. In the video processing context, this is called the line buffer.

If you are using a 9-tap filter in the vertical dimension you will need 9 line buffer memories.

Figure 5.6 shows the resources required to implement a filter. In general you will need memories for each video line store as well as memory for storing the coefficient set. You will also need multipliers for generating the product of the coefficient and pixels, and finally an adder to sum the products.

One way to implement this 2-D scaler (i.e. 2-D filter) is by cascading two 1-D filters, as shown in Figure 5.7 — this is an implementation that is published by Altera for their FPGAs.

The implementation in Figure 5.7 consists of two stages, one for each 1-D filter. In the first stage, the vertical lines of pixels are fed into line delay buffers and then fed to an array of parallel

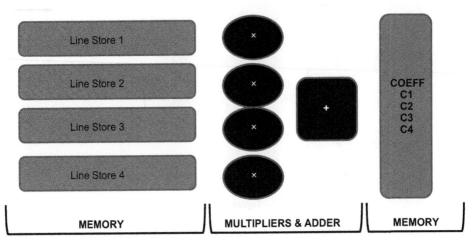

**Figure 5.6.**

multipliers. The outputs of the multipliers are then summed and sent to a "Bit Narrower" which adjusts the size of the output to fit into the number of bits allowed. The second stage has the same basic structure as the first, and filters the "horizontal pixels" output from the first stage to produce the final output pixel.

This structure can be extended to perform $5 \times 5$, $6 \times 6$, or $9 \times 9$ multi-tap scaling. The principles remain the same, but larger kernels will require more FPGA resources. As mentioned previously, each new pixel may use a different set of coefficients depending on its location and will therefore require different coefficient sets.

A programmable logic device with abundant hardware resources, such as an FPGA, is a good platform on which to implement a video scaling engine. In the implementation discussed, four multipliers are needed for scaling in the vertical dimension (one for each column of the scaling kernel), four multipliers are needed for scaling in the horizontal dimension (one for each row of the kernel), and a significant amount of on-chip memory is needed for video line buffers.

Generic DSP architectures, which typically have 1−2 multiply-and-accumulate (MAC) units and significantly lower memory bandwidth, do not have the parallelism for such an implementation. However there are specialized DSPs that have dedicated hardware to implement HD scaling in real-time.

There are various Intellectual Property (IP) providers for video scaling functions in FPGAs. FPGA suppliers provide their own functions as well, which considerably reduces the complexity of implementing the function.

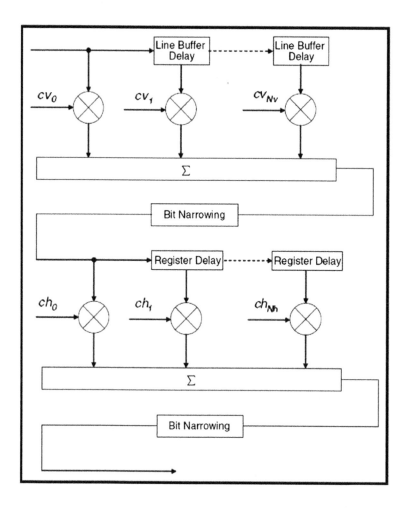

**Figure 5.7.**

These IP functions abstract away all the mathematical details, enabling designers to implement highly complex scaling algorithms in a matter of minutes. Using such functions you can generate a set of scaling coefficients using standard polynomial interpolation algorithms, or use your own custom coefficients.

Figure 5.8 shows the GUI of an IP function provided by Altera for polyphase video scaling. The scaler allows implementation of both simple, nearest neighbor/bilinear scaling, as well as polyphase scaling. For polyphase scaling, the number of vertical and horizontal taps can be selected independently. The number of phases can also be set independently in each dimension. This IP function uses an interpolation algorithm called the Lanczos function to calculate the coefficients.

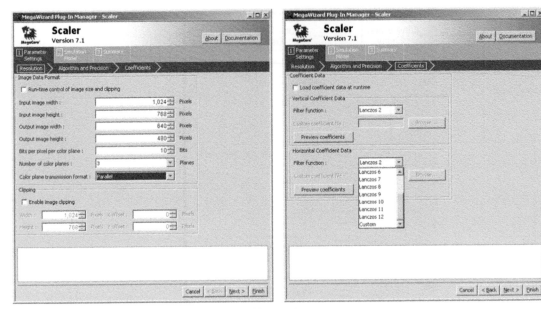

Figure 5.8.

# 5.3 Video Scaling for Different Aspect Ratios

This topic can be confusing, so we will limit the discussion to three common aspect ratios. We will focus on converting everything to the nearly ubiquitous 16:9 aspect ratio found in today's HDTVs.

Aspect ratio is essentially the width of a video frame divided by its height — both expressed in one unit. The easiest way to understand this is to use Excel and consider each cell as one unit. If you highlight four cells in the horizontal direction and three in the vertical direction you get the infamous 4:3 aspect ratio. This was, and is, the aspect ratio of standard TV — an artefact which gives us the annoying black bars on our regular HDTVs. Of course, you can highlight 12 cells in the horizontal direction and nine in the vertical direction and still get the same aspect ratio i.e. 12:9 or 4:3.

All HDTVs have an aspect ratio of 16:9. You can reproduce this in Excel by highlighting 16 cells in the horizontal direction and nine cells in the vertical direction.

Cinematic film, on the other hand, uses an aspect ratio of 1.85:1 (amongst others, but this is common). To replicate this in our Excel example, this ratio translates to 16.65:9. Figure 5.9 shows these different aspect ratios.

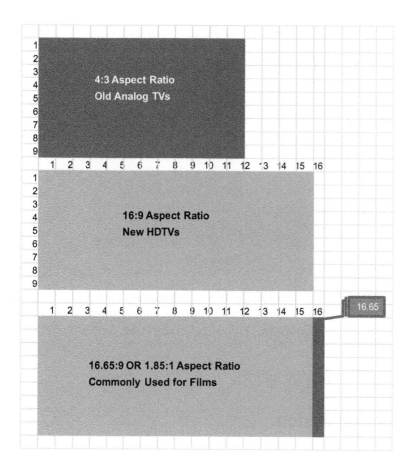

Figure 5.9.

The problem is fitting the 4:3 and the cinematic film aspect ratio onto our HDTV. The simplest way is to take the 4:3 rectangle and fit it on the 16:9 rectangle. When you try this you can immediately understand why we see those black bars on the side. Similarly, when you watch a Blu-ray DVD that is supposed to be HD, you still get bars on the top and bottom. This is because the Blu-ray movie was recorded with the aspect ratio of 1.85:1. Figure 5.10 shows how the black bars appear when we try and fit a different aspect ratio on regular HDTV screen.

The reason these aspect ratios are brought up here is that you can immediately see one of the major uses for video scaling. Most HDTVs have a scaling function which will change the aspect ratio, but most of them employ simplistic scaling algorithms — nearest neighbor or bilinear — and so the resultant image is not very good.

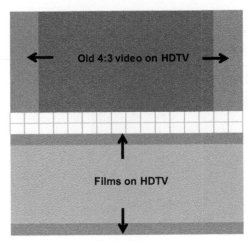

**Figure 5.10.**

Better results occur when the movie makers convert their film aspect ratio to a 16:9 aspect ratio – sometimes known as anamorphic widescreen. This is generally labeled on the DVD.

1.78:1

16:9

WIDESCREEN VERSION

## 5.4 Conclusion

Video scaling is probably one of the most common video processing techniques. It is used in a variety of applications – ranging from HDTV to medical, surveillance and conferencing systems. Done correctly, this is computationally intensive processing which has to be done at the video frame rate. It demands dedicated hardware and/or a fast FPGA to implement at acceptable quality and at fast frame rates.

# VIDEO DEINTERLACING

## CHAPTER OUTLINE

6.1 Basic Deinterlacing Techniques   40
6.2 Motion-Adaptive Deinterlacing: The Basics   43
6.3 Logic Requirements   45
6.4 Cadence Detection   47
   6.4.1 Another feature that is becoming standard in advanced deinterlacers   47
6.5 Conclusion   48

In the days of CRT the video sent to the TV was interlaced (the meaning of which will be discussed later) and the TV monitor on which it was displayed worked perfectly for that format. However, most TVs and monitors are now LCD or Plasma, and these have no ability to interpret interlaced video.

Video deinterlacing techniques were developed to address the problem of legacy interlaced video that was required by old analog televisions.

First we need to understand interlaced video (see Figure 6.1). Consider a video frame that is coming in at 30 fps. One way to represent this would be to break it up into two fields — one field would consist of all the odd numbered rows and one field would consist of all the even numbered rows. Of course, since one frame is now two fields, these would have to be transmitted twice as fast, i.e. at 60 fields per second. This is interlaced video — essentially a succession of 50/60 fields per second, where each field carries only half of the rows that are displayed in each frame of video.

If this sounds convoluted — it is. It was done to support the older analog TVs which were based on CRTs. The electron gun needed time to switch back after "painting" row one and so we needed to skip row two and present it with row three. The detailed operation of a CRT screen is not important for digital video processing. What is important is that we have to deal with this "interlaced" video.

Interlaced video is not suitable for the majority of our monitors today, which paint individual pixels within a single video frame (referred to as progressive).

Digital Video Processing for Engineers. http://dx.doi.org/10.1016/B978-0-12-415760-6.00006-4

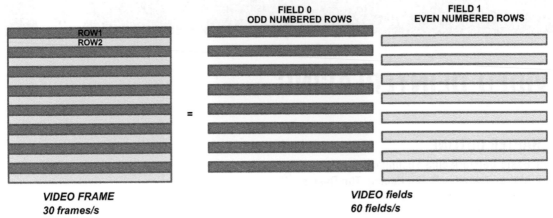

**Figure 6.1.** Interlaced video.

Progressive scanning scans the entire video frame one line at a time and each pixel value within that line is transmitted. Modern monitors put video lines on one at a time in perfect order — ROW 1, ROW 2, ROW 3, ROW 4, etc. Each frame is updated every $1/30^{th}$ of a second — or in other words 30 frames are displayed each second, one after the other.

Our problem surfaces when an interlaced video has to be displayed on a progressive screen. That's where deinterlacing comes into play.

Today, deinterlacing is an important video processing function and is required in many systems. Much video content is available in the interlaced format and almost all of the newer displays — LCD or plasma — require progressive video input.

However, deinterlacing is by nature complex and no algorithm produces a perfect progressive image. Let's look at some basic deinterlacing techniques.

## 6.1 Basic Deinterlacing Techniques

Fundamentally, deinterlacing is the process of taking a stream of interlaced frames and converting it to a stream of progressive frames. One way to do this could be as simple as reversing the Figure 6.1 to Figure 6.2.

In this example we take the two fields and combine them to create one frame. This rather simplistic deinterlacing technique is called "weave" deinterlacing since we are "weaving" two fields to create a single frame. This technique works well

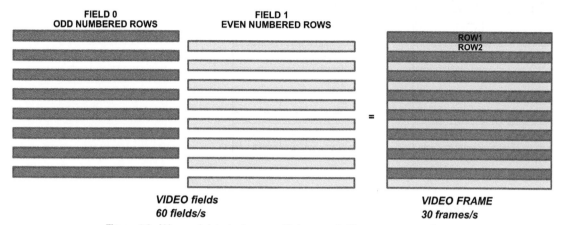

**FIELD 0**
**ODD NUMBERED ROWS**

**FIELD 1**
**EVEN NUMBERED ROWS**

ROW1
ROW2

=

*VIDEO fields*
*60 fields/s*

*VIDEO FRAME*
*30 frames/s*

**Figure 6.2.** Weave deinterlacing: combining two fields to create one frame.

when the two fields are generated from a single progressive video frame.

In practice the two fields are normally offset in time — typically by $1/60$th of a second. So while field zero is captured at time $t_0$ s, field one is captured at time $t_0 + s$. Thus the weave technique is fine if the image does not change in that $1/60$th of the second. But, if the image changes, then the pixels in field zero will not line up with pixels in field one (especially for the portion of the image that changed) — and you will see a jagged edge as shown in Figure 6.3.

Figure 6.3 shows this combing effect — also called mouse teeth — using a simple Excel model. You can immediately and intuitively understand how jagged edges appear where the image has changed and how the resultant image is fine where the image does not change.

Figure 6.4 shows how an image looks when these artifacts appear as a result of weave deinterlacing.

Another form of deinterlacing is called "bob" deinterlacing. where each field becomes its own frame of video. This doubles the resultant frame rate. So an interlaced NTSC clip at 29.97 fps (fields per second) stream becomes a 59.94 fps (fps) progressive. The lines in each field are also doubled as the field becomes a frame — which is why this technique is also sometimes described as spatial line doubling.

Since each field has only half the scan lines of a full frame, interpolation must be used to form the missing scan lines. Interpolation is a fancy term for guessing the value of the line of pixels. In the video scaling chapter we saw how this technique was used to create new pixel values in order to make an image bigger or smaller.

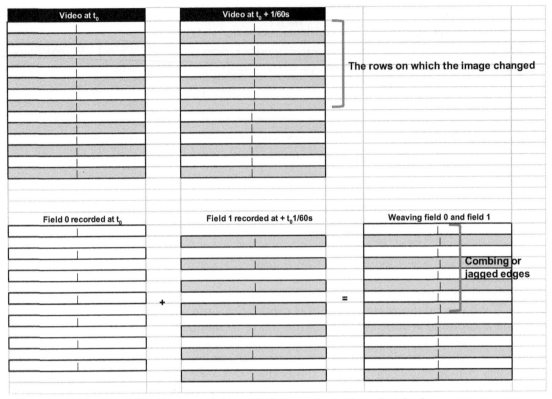

**Figure 6.3.** "Mouse teeth" jagged edges are caused by a changing image.

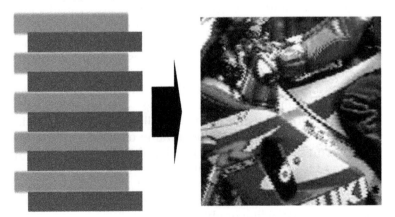

**Figure 6.4.** An image with jagged edges as a result of weave deinterlacing.

Conceptually this deinterlacing technique is the same — however we have to come up with values for the entire line of pixels. The new line can either be just a copy of the previous line (scan-line duplication) or computed as an average of the lines above and below (scan-line interpolation), as shown in Figure 6.5.

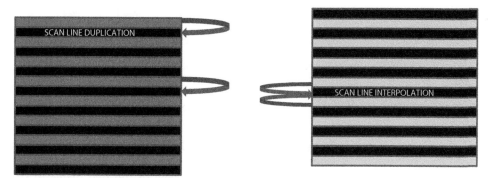

**Figure 6.5.** "Bob" deinterlacing.

Bob deinterlacing provides good results when the image intensity varies smoothly, but it can soften the image because it also reduces the vertical resolution.

Both bob and weave deinterlacing can affect the image quality, especially when there is motion. The bob method can soften the image and the weave method can create jagged images or mouse teeth artifacts. Figure 6.4 contrasts an image generated with the bob technique with one generated with the weave technique. An obvious way to get better quality deinterlacing would be to mix up both the techniques described in the preceding section, after computing whether there is motion between successive frames of video. This technique, which advocates the weave technique for static regions and the bob technique for regions that exhibit motion, is referred to as "motion-adaptive deinterlacing."

## 6.2 Motion-Adaptive Deinterlacing: The Basics

The key to motion-adaptive deinterlacing is to estimate "motion". This is the most computationally intensive task as it assesses the differences of one field to another — and remember there are 60 of these fields passing through in one second.

This is usually done by looking at a window of, for example, $3 \times 3$ on each field. Since field zero contains row one, skips row two and contains row three, this means a $3 \times 3$ window will give you two rows of three pixels each. In field one — since both rows one and row three are skipped — you will only get row two with three pixels (see Figure 6.6).

One way to understand motion adaptive deinterlacing is to understand that you have two options to calculate the value of a missing pixel:

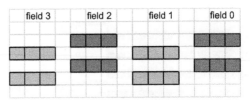

**Figure 6.6.** Motion-adaptive deinterlacing.

Option 1 is to use the pixel value from the following field — i.e. weave deinterlacing. This is illustrated in Figure 6.7.

Option 2 is to use the average of the two pixels in the same field — one above it and one below it or bob deinterlacing (scan line interpolation). This is illustrated in Figure 6.8.

Of course, if there is absolutely no motion, one should go with option 1, but if there is "infinite" motion, field 1 is incompatible and option 2 should be selected. In reality there is always some motion (never infinite) — so we reach for a compromise.

But first let's calculate motion:

Weave the fields to create two frames as shown in Figure 6.9.

Calculate the sum of the nine pixels — pixels are normally represented by 20 to 30 bits.

Take the difference between S0 and S1. This is M → the motion value for this window.

Use the value of M to determine how much you would veer towards option 1 or option 2. The simplest strategy would be to weigh option 1 heavily if M is small and option 2 heavily if M is large i.e. output pixel $=$ (Option 1 value) $\times (1-M) +$ (Option 2 value) $\times$ (M).

The motion value calculated can be used as is or compared to the previous motion value generated. If the previous motion value is higher, then the current motion value is adjusted so that it

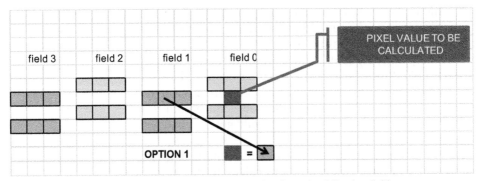

**Figure 6.7.** Option 1: use the pixel value from the following field.

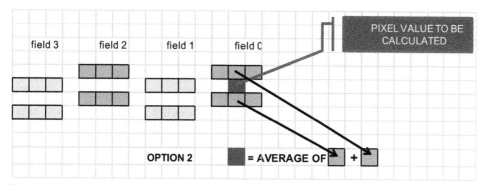

**Figure 6.8.** Option 2: use the average of the two pixels in the same field (one above and one below).

is between the calculated amount and the previous amount. This additional computation is also called "motion bleed" because the motion values from more than one frame in the past are carried over. It is an exponential decay; after a motion, it may take 3 to 10 frames before the weave is again stabilized.

Of course all manner of algorithms can be applied and all of these algorithms — as long as they use a concept of estimated motion — would be labeled "motion adaptive deinterlacing".

## 6.3 Logic Requirements

The bob deinterlacing method with a scan-line duplication algorithm is the simplest and cheapest in terms of logic. Output frames are produced by simply repeating every line in the current field twice. If the output frame rate used is the same as the input frame rate, then half of the input fields are discarded because only the current field is used.

The bob deinterlacing method with a scan-line interpolation algorithm has a slightly higher logic cost than bob with scan-line duplication, but offers significantly better quality. Output frames are produced by filling in the missing lines from the current field with the linear interpolation of the lines above and below. At the top of an F1 field or the bottom of an F0 field where only one line is available, that line is just duplicated. Again, if the output frame rate used is the same as the input frame rate, then half of the input fields are discarded because only the current field is used.

The weave deinterlacing method creates an output frame by filling all of the missing lines in the current field with lines from the previous field. This option gives good results for still parts of an image but unpleasant artifacts in moving parts. The weave algorithm requires external memory to store the fields.

This makes the weave algorithm significantly more expensive in logic elements and external RAM bandwidth than either of the bob algorithms. As mentioned before, the results of the weave algorithm can sometimes be perfect, in the instance where pairs of interlaced fields have been created from original progressive frames or when there is little-to-no motion.

Motion-adaptive deinterlacing provides the best quality but requires greater logic and memory resources. In the simple motion adaptive scenario shown above, we have to store the four fields in external memory, store the value of M (motion) in memory, and dedicate logic for calculating the sum and then the difference between the sums.

Ideally, deinterlacers are implemented in hardware, and FPGAs are used to implement sophisticated high-definition deinterlacers. Memory is the most important hardware resource required to build a highly efficient deinterlacer. This applies to both on-chip memory to store the m × n block of pixels across the different fields (the calculated and previous motion value matrices), as well as the external (generally DDR) memory to store multiple-input video fields and the calculated frames.

Table 6.1 shows the resources used to implement a motion-adaptive deinterlacing algorithm on a PAL video source in Altera Cyclone® III and Stratix® III FPGAs.

The table contrasts the resources used for a motion-adaptive deinterlacing technique with the resources used for a simple weave technique. Notice the drop-off in the amount of memory used even when the weave technique is applied to a higher resolution image.

The single biggest resource that must be carefully considered for implementing motion-adaptive deinterlacing is external memory bandwidth. Fields, motion values and output frames have to be moved into and out of the external memory at the video frame rate.

Thus an important consideration in the design of a deinterlacer is the expected memory bandwidth. To buffer one field of a 480i video source requires 165.7 Mbps:

$(720 \times 240 \text{ pixels / field}) \times (16 \text{ bits / pixel}) \times (59.94 \text{ fields / sec}) = 165 \text{ Mbps}$

The bandwidth doubles for a progressive frame and increases even more for HD video. To calculate the bandwidth, calculate the number of frame accesses that a deinterlacer has to execute and then add the total bandwidth. Compare this to the expected bandwidth of the DDR memory interface, which depends on the throughput and the width of the memory interface.

# 6.4 Cadence Detection

## 6.4.1 Another feature that is becoming standard in advanced deinterlacers

Interlaced video can be even more complex than the transmission of odd and even fields. Motion picture photography is progressive and is based on 24 fps, whereas the NTSC format is 60 fps. The conversion of motion picture photography into interlaced video requires us to convert 24 frames (per second) into 60 fields (per second). Since there is no direct factor, first let's consider what happens if each frame is converted into two fields.

24 frames would convert into 49 fields — not the 60 fields that we are looking for.

One method takes the first frame and converts it to three fields (repeat one field); takes the next frame and converts that to two fields. This is called a 3:2 pull-down technique — or "cadence".

| | | |
|---|---|---|
| Frame 1 | → | 3 Fields |
| Frame 2 | → | 2 Fields |
| Frame 3 | → | 3 Fields |
| Frame 4 | → | 2 Fields |
| Frame 5 | → | 2 Fields |
| *Repeat* | | |

This will give you 60 fields per second — since 12 frames contribute 24 fields and the other 12 frames contribute 36 fields.

Although 24 fps film, and its associated 3:2 video cadence, is the most common format, professional camcorders and various types of video processing use different types of cadences.

For example, Panasonic produced a different cadence for their camcorders. Instead of converting the frames into fields using a repeating 3:2 pattern, the frames are converted into a 2:3:3:2 pattern. The first frame is converted into two fields, the second into three fields, the third into three fields, and the fourth into two fields. It then repeats this pattern for every group of four frames that follows.

Cadence detection is important for deinterlacers because if the correct cadence is not detected video data may be thrown away by the deinterlacer, or processing is done on the wrong field.

For example, if a 3:2 cadence is detected the deinterlacer can suspend its motion adaptive mode, throw away the repeating field, and weave the fields to get perfect deinterlacing with minimal effort.

Deinterlacers must have logic built-in to detect the cadence — this can get even more complex when one part of the frame might have 3:2 cadence, while the other part may be straight interlaced (e.g. a film is inserted in an interlaced video). To detect and correctly deinterlace such a source would require deinterlacers to implement per-pixel cadence detection.

## 6.5 Conclusion

Deinterlacing is an important and common technique used in a range of video processing systems. This is probably not going to change in the near future, because we have to deal with legacy video formats and also legacy monitors.

As with all video processing techniques, deinterlacing can be as complex or as simple as the available computational resources. The simplest techniques are easiest to implement but also produce mediocre quality results. Motion-adaptive deinterlacing, while complex and hardware intensive, provides high quality results.

# 7

# ALPHA BLENDING

## CHAPTER OUTLINE

7.1 Introduction 49
7.2 Concept and Math Behind Alpha Blending 50
7.3 Implementing Alpha Blending in Hardware 51
7.4 Creating a Different Background 51
7.5 Conclusion 52

Alpha blending, or alpha compositing, is a video-processing technique that is used to combine two or more video streams on the same frame. This technique is increasingly used to overlay some graphics on a video, such as the 10-yard line in a football match or to put a different background behind a speaker.

We will learn how this is achieved and what kind of processing is required to overlay images.

## 7.1 Introduction

We will start with two images. Video consists of frames moving at a certain speed (30 fps or 60 fps) and we generally talk about processing within the frame, or spatial processing. When we process across frames it is called temporal processing. Alpha blending is an example of spatial processing.

Each frame is made of pixels, so when we alpha blend two frames we are blending two sets of pixels and coming up with a new pixel in the process. How we blend the two pixels depends on what we want to achieve.

We will start with a simple example: you want to overlay a logo on a frame and you want the logo to appear in the left bottom corner for each and every frame. The simple thing would be to identify all the pixels on the original image that would be overlain by the logo and replace those pixels with the pixels that constitute that logo.

This is simple in principal but hard to do across video frames that are zipping past at 60 fps. What if you did not want a simple overlay, but a translucent logo? Or what if in some video clips you didn't want a logo? All of these options and more can be accommodated in real-time by a silicon architecture that can do fast alpha-blending.

## 7.2 Concept and Math Behind Alpha Blending

First the concept: to mix two frames of video we define a third frame of video called the alpha frame. This frame also has pixels that I will call alpha values. When pixels from frame one are flying through, and the frame that constitutes the logo appears, the hardware creates a composite pixel whose value is defined as:

Composite pixel = pixel from frame 1 × alpha + pixel from frame 2 × (1−alpha)

Think about this simple equation. The alpha frame can be defined as one in all the positions where there is no logo in frame two. And the alpha values are defined as zero in those places where the logo appears.

Composite pixel = pixel from frame one (when the alpha value is one, i.e. where there is no logo)

Composite pixel = pixel from frame two (when the alpha value is zero, i.e. where there is logo)

And that is the math behind alpha blending.

There are many advantages to defining the blending of video in this manner. For example, if you want the logo to be translucent, you define the alpha value as 0.5 wherever the logo appears. Therefore the value of the composite pixel in places where the logo appears becomes:

Composite pixel = pixel from frame 1 × 0.5 + pixel from frame 2 × 0.5

This could be described as half-and-half or, in other words, the logo is translucent.

To take another example, you may want to blend three frames of video. You define two alpha frames: the first alpha frame is used to blend frame one and frame two: the second alpha frame blends the newly created composite frame with frame three. And now you have a blended frame.

If you wish to remove the logo from the frames for the next five minutes, you would leave the hardware in place and just update the alpha frame for the next five minutes. That would wipe out the logo, and after five minutes the logo would come back.

Many effects can be achieved with this simple concept.

## 7.3 Implementing Alpha Blending in Hardware

Although it is simple in concept, implementing alpha blending in real-time requires fast, dedicated hardware. Let's take a look at the equation above — written in a short form here:

$$Ci = p1i(\alpha) + p2i(1-\alpha)$$

This means you need two multipliers and one adder. As implementing a multiplier in hardware is expensive, you can rearrange your equation as shown in Figure 7.1 to minimize the multipliers.

In Figure 7.1:

Composite pixel can be computed as:

$$c_r = \alpha\, f_r + (1 - \alpha)\, b_r$$
$$c_g = \alpha\, f_g + (1 - \alpha)\, b_g$$
$$c_b = \alpha\, f_b + (1 - \alpha)\, b_b$$

This can also be written as:

$$c_r = b_r + \alpha\, (f_r - b_r)$$
$$c_g = b_g + \alpha\, (f_g - b_g)$$
$$c_b = b_b + \alpha\, (f_b - b_b)$$

One multiply operation vs. two

**Figure 7.1.** The equations for alpha blending.

C is the composite pixel value.
F is the foreground pixel — we were referring to it as the pixel from frame one.
B is the background pixel — we were referring to it as the pixel from frame two.
Figure 7.1 also illustrates that when we talk about creating a composite pixel, the calculations have to be done separately for each color component. We do the same calculation for red, green and blue. Which means not two multipliers, but six multipliers for one pixel...

A simple rearrangement of the equation as shown lets us use only one multiplier and two adders for each color component. Which means three multipliers for the pixel...

As adders are cheaper in silicon than multipliers this is the route we would choose. Figure 7.2 shows the hardware required to calculate the three color components of the composite pixel.

## 7.4 Creating a Different Background

It is common to see a background, behind a newscaster or a weatherman for example, that is different to the actual physical

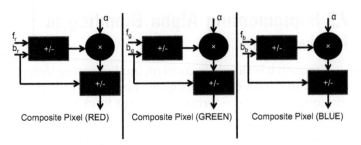

**Figure 7.2.** The hardware to implement the alpha blending equations.

background behind that person. As the person moves, the background adjusts perfectly. This effect uses alpha blending principles.

First the subject sits or stands in front of a background that is a different color (such as bright green) to the subject's shirt, suit or face.

The background on which we wish to superimpose the subject is shot separately. Then all we have to do is alpha blend – with a slight change. Each pixel is compared to the green value. If the pixel is green, then by adjusting the alpha value the desired background pixel wins and replaces the green pixel. If the pixel is not green then we can safely assume that it is some portion of the subject's anatomy, and it is left alone. The background pixel loses and the subject's pixel wins.

By virtue of this approach the composite frame is adjusted so that the background changes around the subject smoothly, even in motion.

This is but one example of the very useful type of video processing that can be done using alpha blending.

## 7.5 Conclusion

Readers beginning with video processing must understand that the alpha blending function has uses that ranges from simple graphics overlay to creating special effects with multiple video streams. But in spite of how different these effects may look – the fact remains that what is required to be done is to generate a composite pixel – which is a function of two different pixels. In effect the underlying function – alpha blending – is the same.

Alpha blending, though mathematically a simple concept requires fast hardware since the processing has to be done at pixel rates. As we have seen earlier these pixel rates are very high especially for HD video. The applications of alpha-blending are many and with the advent of low-cost FPGAs that can handle HD pixel rates, this computationally intensive function can be provided in a cost-effective manner.

# 8

# SENSOR PROCESSING FOR IMAGE SENSORS

## CHAPTER OUTLINE

8.1 CMOS Sensor Basics   54
8.2 A Simplistic HW Implementation of Bayer Demosaicing   56
8.3 Sensor Processing in Military Electro-optical Infrared Systems   56
8.4 Conclusion   58

There are two primary types of sensors that are used to capture video data — CCD sensors and CMOS sensors. A significant difference between these two technologies is how the video data is read out. For CCD sensors the charge is shifted out whereas for CMOS sensors the charge or voltage flows through column and row decoders, very much like a digital memory architecture.

CMOS sensors are becoming prevalent for a variety of reasons — not least because of their proliferation within consumer devices. This not only drives up the quality but also drives down the price. The latest iPhone has an 8 MP CMOS sensor — the kind that was found only in the most expensive digital SLR cameras just five years ago.

CMOS sensors are rapidly becoming the *de facto* standard when it comes to digital image capture. The output of an image sensor is not in a standard video format that can be processed: it cannot be resized (scaled), deinterlaced, or composited for example.

The sensor image must first go through a processing pipeline that may include a range of functions such as pixel correction, noise reduction, Bayer-to-RGB conversion, color correction, gamma correction, and other functions, as shown in Figure 8.1.

Given the wide variety of image sensors used, and the range of application requirements, each processing pipeline will be unique. In many cases hardware customization is required to produce an optimum quality image.

Digital Video Processing for Engineers. http://dx.doi.org/10.1016/B978-0-12-415760-6.00008-8

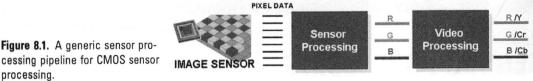

**Figure 8.1.** A generic sensor processing pipeline for CMOS sensor processing.

## 8.1 CMOS Sensor Basics

A CMOS sensor will generally consist of a grid of pixel sensors, each containing a photo-detector and an active amplifier.

CMOS sensors record the image data in grayscale. Color information is gathered by applying a color filter over the pixel grid. Such a color filter array (CFA) allows only light of a given primary color (R, G, or B) to pass through — everything else is absorbed, as shown in Figure 8.2.

This means each pixel sensor collects information about only one color. This is important to recognise as the data for each pixel is not composed of three colors (RGB), but of a single color. The other colors need to be "guessed" (interpolated) by the electronics behind the CMOS sensor.

There are many types of color filter arrays — the most common was invented by Dr. Bayer of Eastman Kodak. This color filter array — or the Bayer mosaic as it is more commonly referred to — utilizes a filter pattern that is half green, a quarter red and a quarter blue. This is based on the knowledge that the human eye is more sensitive to green.

The resultant image data does not have all three color components for each pixel. For some pixels we have only red data, for some only green and for others we have only blue. To do any meaningful video processing each pixel must have all the color plane data. So we must process this Bayer image to get three color planes for each pixel.

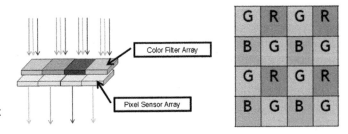

**Figure 8.2.** The Bayer color filter array that gives more green than red and blue data.

This processing is called the Bayer to RGB conversion, or Bayer demosaicing, and it is the key standard function required in a sensor processing-signal chain. This is also the most complex when implemented using a multi-tap filter.

Referring back to our discussion of video scaling, when you don't have the pixel data, you must create it, not randomly through guesswork, but in an informed manner by looking at the data from the neighboring pixels (see Chapter 5 for details).

The simplest way would be to just copy the color plane values from the pixel that is the nearest and hope for the best. A second way would be to take an average of the neighboring pixels (see Figure 8.3).

In this figure focus on the center pixel: we will be looking at ways to calculate the missing color information for this pixel.

The figure shows four possible cases. In the first two cases the center pixel has only a green value and in the next two cases the center value has either the blue color information or the red color information.

In (a) and (b), the red and the blue color planes can be calculated by taking the average of the neighboring pixels. In (c) and (d), since there are four neighboring green pixels and four neighboring blue pixels, the missing color plane can be calculated as:

$$(G1 + G2 + G3 + G4) / 4 \ OR \ (B1 + B2 + B3 + B4) / 4$$

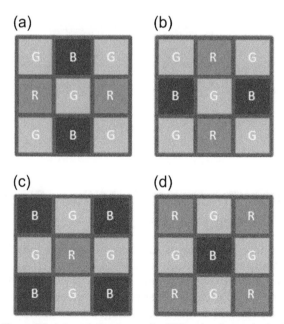

**Figure 8.3.** Interpolating to get the RGB value for each pixel.

This is called the bilinear interpolation technique. This is exactly the same as the video scaling techniques discussed earlier.

As in video scaling, the complexity in the interpolation technique can be increased to $5 \times 5$, $6 \times 6$ or to any arbitrary $n \times n$ level. The disadvantage is the additional hardware resources that are employed to implement this algorithm.

## 8.2 A Simplistic HW Implementation of Bayer Demosaicing

Since these algorithms have to be run at the pixel rate for HD (high-definition) resolutions, the pixel rate is very high and so FPGAs are used to implement custom demosaicing algorithms. The logic inside the FPGA is structured as shown in Figure 8.4.

The Bayer data from the camera is fed into custom line buffers built inside the FPGA. These line buffers are "n" wide, where "n" is the horizontal resolution of the sensor (how many pixels per line the sensor can provide).

The number of the line buffers that you need depends upon your Bayer demosaicing algorithm. A bilinear implementation may require just two line buffers, but a complex $5 \times 5$ interpolation would need more line buffer resources.

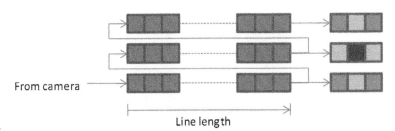

From camera

Line length

**Figure 8.4.** Line buffer implementation for Bayer Demosaicing.

## 8.3 Sensor Processing in Military Electro-optical Infrared Systems

Military imaging systems are becoming increasingly sophisticated, incorporating multiple advanced sensors ranging from thermal infrared to visible, and even to UV (ultraviolet) focal planes. Not only do these sensor outputs need to be corrected, interpolated and so forth, often images from multiple sensors must be fused, overlaid, and further processed for local display and/or for transmission.

Figure 8.5 shows a high-level block diagram of a typical signal chain implemented in an electro-optical infrared (EOIR) system. As shown, the processed image is compressed many times (usually using lossless techniques) before being transmitted over a communications link.

The first group of algorithms shown is responsible for the configuration and operation of the image sensor (also called focal plane array (FPA)). These algorithms include the generation of video timing and control signals for exposure control, readout logic, and synchronization.

Once this is completed, the pixel streams are processed by a second group of algorithms that addresses the imperfections of the focal plane. Functions such as non-uniformity correction, defective pixel replacement, noise filtering, and pixel binning may also be used to improve image quality. For a color-capable focal plane, demosaicing may be performed. The corrected video stream is then processed to implement functions such as automatic gain and exposure control, wide dynamic range (WDR) processing, white balancing and gamma correction.

In addition, FPGA-based camera cores are able to implement video processing algorithms that further enhance the output video. These processing stages may include functions such as image scaling (digital zoom), (de)warping, windowing, electronic image stabilization, super-resolution, external video overlay, image fusion, and on-screen display. In some cases, the captured and processed video stream may need to be compressed before it is transmitted to a control station.

The EOIR system implements high-quality sensor control and image processing within a tightly constrained power budget, yet

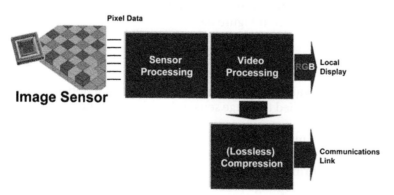

**Figure 8.5.** Typical top level structure of an EOIR system.

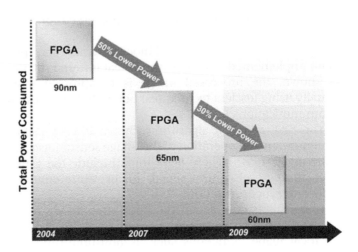

**Figure 8.6.** Power consumption trend of today's leading FPGAs.

retaining the programmability required for last-minute specification changes and field upgradeability.

Combining exceptional image quality with low power consumption is the key challenge when designing EOIR systems. For hand-held and wearable systems, such as night-vision goggles (NVGs) or weapon sights, the critical specification is often the number of hours a unit can run on AA batteries.

Low-power FPGAs are the platform of choice for almost all state-of-the-art EOIR systems because they meet the need for programmability, real-time video processing performance, and low power consumption. In practice, each successful generation of low-power FPGAs have achieved both lower static and dynamic power consumption by utilizing a combination of architectural enhancements and lower core voltages. As the process technology continues to march downwards, the average power consumed by these FPGAs has been dropping by up to 50% and 30% (as shown in Figure 8.6).

## 8.4 Conclusion

There are various other sensor processing functions that can be, and are, applied to image sensor data. Some of these functions include:

Digital zoom/binning.
Noise filtering.
Non-uniformity correction.
Wide dynamic range processing.
Local-area adaptive contrast enhancement.

Pixel-level adaptive image fusion.

Electronic image stabilization.

Super-resolution.

Motion detection and multi-target tracking.

It is beyond the scope of an introductory text such as this to delve into all of these functions. As the image sensors that many of you will work with will be a CMOS sensor with a Bayer output format, it is important to understand how this data is processed so that full three color plane information for each pixel is created.

High-performance sensor processing at very low power is increasingly important in various military applications and we have seen that this kind of processing requires the inherent parallel architecture of FPGAs.

# 9

# VIDEO INTERFACES

## CHAPTER OUTLINE

9.1 SDI   61
9.2 Display Port   61
9.3 HDMI   63
9.4 DVI   63
9.5 VGA   64
9.6 CVBS   64
9.7 S-Video   64
9.8 Component Video   65

There are several common video interfaces, which are used both in the broadcast industry and among consumer products. These are briefly described below:

## 9.1 SDI

This is a broadcast industry standard, used to interconnect a variety of professional equipment in broadcast studios and mobile video processing centers (like big truck trailers seen at major sports events). SDI stands for "serial data interface", which is not very descriptive. It is an analog signal, modulated with digital information. This is usually connected using a co-axial cable, normally BNC type as in Figure 9.1.

It is able to carry all of the data rates listed in Table 9.1 and dynamically switch between them. Most FPGAs and broadcast ASICs can interface directly with SDI signals.

## 9.2 Display Port

Display port is a next-generation video-display interface. It uses from one to four lanes of video data – each lane is a single wire, with the clock embedded in the serialized data. Each lane operates at 1.6 Gbps, 2.7 Gbps, and in latest versions, 5.4 Gbps. It also features an auxiliary channel for low-rate two-way configuration and status information between the video source and

Digital Video Processing for Engineers. http://dx.doi.org/10.1016/B978-0-12-415760-6.00009-X

**Figure 9.1.** SDI cable.

# Table 9.1

| Image Size | Frame Size | Color Plane Format at 60 fps | Bit/s transfer rate |
|---|---|---|---|
| 1080p × 1920 | 1125 × 2200 | 4:2:2 YCrCb | 2200 × 1125 × 20 × 60 = 2.97 Gbps |
| 1080i × 1920 | 1125 × 2200 | 4:2:2 YCrCb | 2200 × 1125 × 20 × 60 × 0.5 = 1.485 Gbps |
| 720p × 1280 | 750 × 1650 | 4:2:2 YCrCb | 1650 × 750 × 20 × 60 = 1.485 Gbps |
| 480i × 720 | 525 × 858 | 4:2:2 YCrCb | 858 × 525 × 20 × 60 × 0.5 = 270 Mbps |

display. It can also carry DC power over the connectors. This connection type is common in newer consumer equipment. The display port protocol also contains video encryption features for security, known as HDPC (high definition digital content protection).

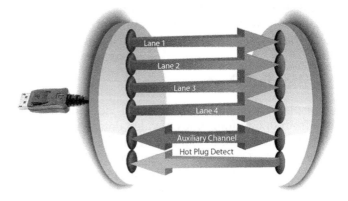

**Figure 9.2.** Display port.

## 9.3 HDMI

HDMI or High-Definition Multimedia Interface carries digital audio/video data between video sources and display devices, such as monitors or big-screen TVs. HDMI signals are electrically compatible with a DVI signal, so conversion is necessary. HDMI carries various resolutions, both compressed and uncompressed, including high-definition video. Different stereo audio formats are also defined. Various capabilities and data rates are available depending upon the version of HDMI supported.

**Figure 9.3.** HDMI connector.

## 9.4 DVI

DVI, or Digital Video Interface, is a connection type common on newer computer monitors. It is a multi-pin connector carrying separated RGB, digitized video information at the desired frame resolution.

**Figure 9.4.** DVI connector.

## 9.5 VGA

This is used to connect most computer monitors, known as VGA. It carries analog RGB signals using the familiar multi-pin "sub-D" connector located on the back, or side, of all computers. This connects the computer to monitors, or laptops to projectors for whole room viewing. This connector carries separated RGB video information at the desired frame resolution, using analog signals.

**Figure 9.5.** VGA monitor connector.

## 9.6 CVBS

The Composite Video Blanking and Sync cable is the basic yellow one often used to connect televisions and VCRs or DVDs together. It carries a SD 4:2:2 YCrCb combined analog video signal on a low-cost coax "patch cable".

**Figure 9.6.** Composite cable.

## 9.7 S-Video

This is commonly used to connect consumer home-theater equipment such as flat panels, televisions and VCRs or DVDs together. It carries a 4:2:2 YCrCb signals in separate wires over a single multi-pin connector, using a shielded cable. It is higher quality than CVBS.

**Figure 9.7.** S-video cable.

# 9.8 Component Video

This is another common way to connect consumer home-theater equipment such as flat panels, televisions and VCRs or DVDs together. It carries a 4:2:2 YCrCb analog signal in separate form over a three-coax patch cable. Often the connectors are labeled as Y, $P_B$, $P_R$. It is higher quality than S-Video due to separate cables.

**Figure 9.8.** Component video cables.

# VIDEO ROTATION

## CHAPTER OUTLINE
10.1 Interpolation   68

Video images can be rotated, as well as scaled. Common applications involve surveillance, heads-up displays, military and industrial usages. Images are rotated about a reference point in the image. This can be seen in the image rotation below. Depending upon the location of the reference point, the rotation may also cause the image to move to a different location.

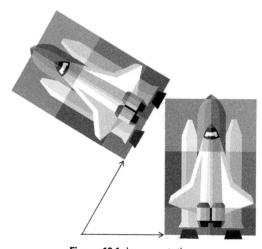

**Figure 10.1** Image rotation.

The mathematical relationship is quite simple. To perform a rotation of $\theta$, with a normalized reference point (0,0), a pixel at location $(x,y)$ is translated to $(x', y')$.

$$x' = \cos(\theta) \times x - \sin(\theta) \times y$$
$$y = \sin(\theta) \times x + \sin(\theta) \times y$$

This can be expressed in matrix form as:

Digital Video Processing for Engineers. http://dx.doi.org/10.1016/B978-0-12-415760-6.00010-6

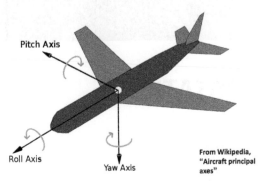

Pitch Axis

Roll Axis

Yaw Axis

**Figure 10.2** Three-dimensional rotation.

$$\begin{bmatrix} x' \\ y' \end{bmatrix} = \begin{bmatrix} \cos\theta & -\sin\theta \\ \sin\theta & \cos\theta \end{bmatrix} \begin{bmatrix} x \\ y \end{bmatrix}$$

For rotation about an arbitrary reference point $X_0$, $Y_0$, the following substitutions are used:

$$(x - X_0) = \cos(\theta) \times (x - X_0) - \sin(\theta) \times (y - Y_0)$$
$$(y' - Y_0) = \sin(\theta) \times (x - X_0) + \sin(\theta) \times (y - Y_0)$$

Computing rotations involves multiplications, plus image blending, to overwrite the original pixels with the rotated image.

Rotation for video and imaging applications is considered only for two dimensions. More sophisticated three-dimensional rotations are performed in visual applications such as computer generated graphics, design visualization software and other uses involving depth.

One way to think about this is in the aircraft terms of yaw, roll and pitch.

This motion can be represented using a $3 \times 3$ matrix, which will rotate a point $x$, $y$, $z$ in three-dimensional space to a new location $x'$, $y'$, $z'$.

$$\mathbf{A}\begin{pmatrix} x \\ y \\ z \end{pmatrix} = \begin{pmatrix} a & b & c \\ d & e & f \\ g & h & i \end{pmatrix}\begin{pmatrix} x \\ y \\ z \end{pmatrix} = \begin{pmatrix} x' \\ y' \\ z' \end{pmatrix}$$

# 10.1 Interpolation

When rotating images, the new location may not be a valid pixel location. It could land mid-way between pixel locations, or in any other arbitrary location. In this case, better quality can be achieved by interpolation, or calculating the actual values at the valid pixel locations. This is achieved by interpolating between nearby rotated pixels that do not line up with the pixel grid locations.

# ENTROPY, PREDICTIVE CODING AND QUANTIZATION

## CHAPTER OUTLINE

11.1 Entropy   69
11.2 Huffman Coding   71
11.3 Markov Source   72
11.4 Predictive Coding   73
11.5 Differential Encoding   74
11.6 Lossless Compression   75
11.7 Quantization   76
11.8 Decibels   79

In this chapter, we will discuss some of the basic concepts of data compression, including video and image compression. Until now we have considered only uncompressed video formats, such as RGB or YCrCb, where each pixel is individually represented (although this is not strictly true for 4:2:2 or 4:2:0 forms of YCrCb). However, high levels of compression are possible with little loss of video quality. Reducing the data needed to represent an individual image or a sequence of video frames is very important when considering how much storage is needed on a camera flash chip or computer hard disk, or the bandwidth needed to transport cable or satellite television, or stream video to a computer or handheld wireless device.

## 11.1 Entropy

We will start with the concept of entropy. Some readers may recall from studies in thermodynamics or physics that entropy is a measure of the disorderliness of a system. Further, the second law of thermodynamics states that in a closed system entropy can only increase, and never decrease. In the study of compression, and also the related field of error correction, entropy can be thought of as the measure of unpredictability. Entropy can be applied to a set of digital data.

Digital Video Processing for Engineers. http://dx.doi.org/10.1016/B978-0-12-415760-6.00011-8

The less predictable a set of digital data is, the more information it carries. Here is a simple example: assume that a bit can be equally likely to be either zero or one. By definition, this will be one bit of data information. Now assume that this bit is known to be one with 100% certainty. This will carry no information, because the outcome is predetermined. This relationship can be generalized by:

Info of outcome $= \log_2$ (1 / probability of outcome) $= -\log_2$ (probability of outcome)

Let's look at another example. If there is a four outcome event, with equal probability of $outcome_1$, $outcome_2$, $outcome_3$, or $outcome_4$:

Outcome 1: Probability $= 0.25$, encode as 00.
Outcome 2: Probability $= 0.25$, encode as 01.
Outcome 3: Probability $= 0.25$, encode as 10.
Outcome 4: Probability $= 0.25$, encode as 11.

The entropy can be defined as the sum of the probabilities of each outcome multiplied by the information conveyed by that outcome.

Entropy $=$ prob $(outcome_1) \times$ info $(outcome_1) +$ prob $(outcome_2) \times$ info $(outcome_2) + \dots$ prob $(outcome_n) \times$ info $(outcome_n)$

For our simple example:

Entropy $= 0.25 \times \log_2(1 / 0.25) + 0.25 \times \log_2(1 / 0.25) + 0.25 \times \log_2(1 / 0.25) + 0.25 \times \log_2(1 / 0.25) = 2$ bits

This is intuitive — two bits is normally what would be used to convey one of four possible outcomes.

In general, the entropy is the highest when the outcomes are equally probable, and therefore totally random. When this is not the case, and the outcomes are not random, the entropy is lower, and it may be possible to take advantage of this and reduce the number of bits to represent the data sequence.

If the probabilities are not equal, for example:

Outcome 1: Probability $= 0.5$, encode as 00.
Outcome 2: Probability $= 0.25$, encode as 01.
Outcome 3: Probability $= 0.125$, encode as 10.
Outcome 4: Probability $= 0.125$, encode as 11.

Entropy $= 0.5 \times \log_2(1 / 0.5) + 0.25 \times \log_2(1 / 0.25) + 0.125 \times \log_2(1 / 0.125) + 0.125 \times \log_2(1 / 0.125) = 1.75$ bits

## 11.2 Huffman Coding

Since the entropy is less than two bits, then, in theory, we should be able to convey this information in less than two bits. What if we encoded these events differently as below:

Outcome 1: Probability $= 0.5$, encode as 0.

Outcome 2: Probability $= 0.25$, encode as 10.

Outcome 3: Probability $= 0.125$, encode as 110.

Outcome 4: Probability $= 0.125$, encode as 111.

For half the time we would present the data with one bit, one quarter of the time with two bits and one quarter of the time with three bits.

Average number of bits $= 0.5 \times 1 + 0.25 \times 2 + 0.125 \times 3 + 0.125 \times 3 = 1.75$

The nice thing about this encoding is that the bits can be put together into a continuous bit stream, and unambiguously decoded.

010110111110100....

The bit stream above can only represent one possible outcome sequence.

$outcome_1$, $outcome_2$, $outcome_3$, $outcome_4$, $outcome_3$, $outcome_2$, $outcome_1$...

An algorithm known as Huffman coding works in this manner, by assigning the shortest codewords (the bit-word that each outcome is mapped to, or encoded) to the events of highest probability. For example, Huffman codes are used in JPEG-based image compression. Originally, this concept was used over 150 years ago in Morse code for telegraphs, where each letter of the alphabet is mapped to a series of short dots and long dashes.

Common letters:

E ·

I ··

T -

Uncommon letters:

X -···

Y -·−

Z −··

In Morse code, the higher probability letters are encoded in a few dots, or a single dash. The lower probability letters use several dashes and dots. This minimized, on average, the amount of time required by the operator to send a telegraph message.

# 11.3 Markov Source

Further opportunities for optimization arise when the probability of each successive outcome or symbol is dependent upon previous outcomes. An obvious example is that if the $n^{th}$ letter is a "q", you can be pretty sure the next letter will be "u". Another example is that if the $n^{th}$ letter is a "t", this raises the probability that the following letter will be an "h". Data sources that have this kind of dependency relationship between successive symbols are known as Markov sources. This can lead to more sophisticated encoding schemes. Common groups of letters with high probabilities can be mapped to specific codewords. Examples are the letter pairs "st" and "tr", or "the".

This can occur when we map the probabilities of a given letter based on the few letters preceding. In essence, we are making predictions based on the probability of certain letter groups appearing in any given construct of the English language. Of course, different languages, even if using the same alphabet, will have a different set of multi-letter probabilities, and therefore different codeword mappings. An easy example in the United States is the sequence of letters: HAWAI_. Out of 26 possibilities, it is very likely the next letter is an "I".

In a first order Markov source, a given symbol's probability is dependent upon the previous symbol. In a second order Markov source, a given symbol's probability is dependent upon the previous two symbols, and so forth. The average entropy of a symbol tends to decrease as the order of the Markov source increases. The complexity of the system also increases as the order of the Markov source increases.

A first order binary Markov source is described in Figure 11.1. The transition probabilities depend only on the current state.

In this case, the entropy can be found as the weighted sum of the conditional entropies corresponding to the transitional probabilities of the state diagram. The probability of a zero given the previous state is a one is described as $P(0|1)$.

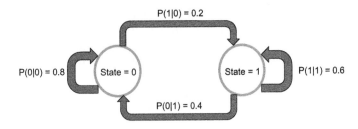

**Figure 11.1.** Markov diagram.

The probability of each state can be found by solving the probability equations:

$$P(0) = P(0) \times P(0|0) + P(1) \times P(0|1) = P(0) \times 0.8 + P(1) \times 0.4$$

Both methods can output a zero (meaning to arrive at state $= 0$ circle), either starting from state zero or one.

$$P(1) = P(1) \times P(1|1) + P(0) \times P(1|0) = P(1) \times 0.6 + P(0) \times 0.2$$

Both methods can output a one (meaning to arrive at state $= 0$ circle), either starting from state zero or one.

By definition $P(0) + P(1) = 1$

From this, we can solve for $P(0)$ and $P(1)$:

$$P(0) = \{2 / 3\}$$
$$P(1) = \{1 / 3\}$$

Now we can solve for the entropy associated with each state circle.

For state zero and state one:

Entropy$_0 = 0.8 \times \log_2(1 / 0.8) + 0.2 \times \log_2(1 / 0.2) = 0.258 + 0.464 = 0.722$ bits

Entropy$_1 = 0.6 \times \log_2(1 / 0.6) + 0.4 \times \log_2(1 / 0.4) = 0.442 + 0.529 = 0.971$ bits

The entropy of the Markov source or system is then given by:

$P(0) \times$ Entropy$_0 + P(1) \times$ Entropy$_1 = 1 / 3 \times 0.722 + 2 / 3 \times 0.971 = 0.888$ bits

## 11.4 Predictive Coding

Finding the sequence "HAWAII" is a form of predictive coding. The probabilities of any Markov source, such as language, can be mapped in multi-letter sequences with an associated probability. These in turn can be encoded using Huffman coding methods, to produce a more efficient representation (fewer number of bits) of the outcome sequence.

This same idea can be applied to images. As we have previously seen, video images are built line by line, from top to bottom, and pixel by pixel, from left to right. Therefore, for a given pixel, the pixels above and to the left are available to help predict the next pixel. For our purposes, let us assume an RGB video frame, with eight bits per pixel and color. Each color can have a value from 0 to 255.

The unknown value pixel, for each color, can be predicted from the pixels immediately left, immediately above, and diagonally above. Three simple predictors are given in Figure 11.2.

The entire frame could be iteratively predicted from just three initial pixels, but this is not likely to be a very useful prediction.

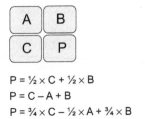

$P = \frac{1}{2} \times C + \frac{1}{2} \times B$

$P = C - A + B$

$P = \frac{3}{4} \times C - \frac{1}{2} \times A + \frac{3}{4} \times B$

**Figure 11.2.** Three examples of video predictors.

The usefulness of these predictors becomes apparent when used in conjunction with differential encoding.

## 11.5 Differential Encoding

Suppose we take the actual pixel value and subtract the predicted pixel value. This is differential encoding, and this difference is used to represent the pixel value for that location. For example, to store a video frame we would perform the differential encoding process, and store the results.

To restore the original video data for display, we would simply need to compute the predicted pixel value (using the previous pixel data) and then add to the stored differential encoded value. This is the original pixel data. (The three pixels on the upper left corner of the frame are stored in their original representation, and are used to create the initial predicted pixel data during the restoration process).

All of this sounds unnecessarily complicated, so what is the purpose here? The differential encoder outputs are most likely to be much smaller values than the original pixel data, due to the correlation between nearby pixels. Since statistically the differentially encoded data is just representing the errors of the predictor, this signal is likely to have the great majority of values concentrated around zero, or to have a very compact histogram. In contrast, the original pixel values are likely to span the entire color value space, and therefore are a high entropy data source with an equal or uniform probability distribution.

Differential encoding sends less information than if the video frame was simply sent pixel by pixel across the whole frame. The entropy of the data stream has been reduced through use of differential encoding, as the correlation between adjacent pixels has been largely eliminated. Without differential encoding, we have no information as to the type of video frames that will be processed, the initial pixel value outcomes are assumed to be equally distributed, meaning each of the 256 possible values has a probability of 1/256, and an entropy of eight bits per pixel. The possible outcomes of the differential encoder tend to be much more probable for small values (due to the correlation to nearby pixels) and much less likely for larger values. As we saw in our simple example above, when the probabilities are not evenly distributed, the entropy is lower. Lower entropy means less information. Therefore, on average, significantly fewer bits are required to represent the image, and the only cost is increased complexity due to the predictor and differential encoding computations.

Let's assume that averaged over a complete video frame, the probabilities work out thus:

Pixel color value Probability $= 1/256$ for values in range of 0 to 255

And that after differential encoding, the probability distribution comes out as:

| Differential color | Probability $= 1/16$ | for value equal to 0 |
|---|---|---|
| Differential color | Probability $= 1/25$ | for values in the range $-8$ to $-1$, 1 to 8 |
| Differential color | Probability $= 1/400$ | for values in the range $-32$ to $-9$, 9 to 32 |
| Differential color | Probability $= 1/5575$ | for values in the range $-255$ to $-33$, 33 to 255 |

We earlier defined entropy as:

Entropy $=$ prob (outcome$_1$) $\times$ info (outcome$_1$) $+$ prob (outcome$_2$) $\times$ info (outcome$_2$) $+ \ldots$ prob (outcome$_n$) $\times$ info (outcome$_n$)

with info of outcome $= \log_2$ (1 / probability of that outcome)

In the first case, we have 256 outcomes, each of probability 1/256:

Entropy $= 256 \times (1 / 256) \times \log_2 (1 / (1 / 256)) = 8$ bits

In the second case, we have 511 possible outcomes, with one of four probabilities:

Entropy $= 1 \times (1 / 16) \times \log_2 (1 / (1 / 16)) + 16 \times (1 / 25) \times \log_2 (1 / (1 / 25)) + 48 \times (1 / 400) \times \log_2 (1 / (1 / 400)) + 446 \times (1 / 5575) \times \log_2 (1 / (1 / 5575)) = 0.250 + 2.972 + 1.037 + 0.996 = 5.255$ bits

## 11.6 Lossless Compression

The entropy has been significantly reduced through the use of the differential coder. While this step function probability distribution is obviously contrived, typical entropy values for various differential encoders across actual video frame data tend to be in the range of four to five ½ bits. Huffman coding, or similar mapping of the values to bit codewords, can reduce the number of bits used to transmit each color plane pixel to ~5 bits compared to the original eight bits. This has been achieved without any loss of video information, meaning the reconstructed video is identical to the original. This is known as lossless compression, as there is no loss of information in the compression (encoding) and decompression (decoding) process.

Much higher degrees of compression are possible if we are willing to accept some level of video information loss, which can result in video quality degradation. This is known as lossy compression. Note that with lossy compression, each time the video is compressed and decompressed, some information is lost, and there will a resultant impact on video quality. The trick is to achieve high levels of compression without noticeable video degradation.

## 11.7 Quantization

Many of the techniques used in compression, such as the DCT, Huffman coding and predictive coding, are fully reversible with no loss in video quality. Quantization is often the principal area of compression where information is irretrievably lost, and consequently decoded video will suffer quality degradation. With care, this degradation can be almost imperceptible to the viewer.

Quantization occurs when a signal with many or infinite number of values must be mapped into a finite set of values. In digital signal processing, signals are presented in binary numbers, with $2^n$ possible values mapping into an n bit representation.

For example, suppose we want to present the range $-1$ to $+1$ (well, almost $+1$) using an 8-bit fixed point number. With $2^n$ or 256 possible values to map to across this range, the step size is 1 / 128, which is 0.0078125. Let's say the signal has an actual value of 0.5030. How closely can this value be represented? What if the signal is 1/10 the level of the first sample, or 0.0503? And again, consider a signal with value 1 / 10 the level as the second sample, at 0.00503. Below is a table showing the closest representation just above and below each of these signal levels, and the resultant error in the 8-bit representation of the actual signal to sampled signal value.

The actual error level remains more or less in the same range over the different signal ranges. This error level will fluctuate, depending upon the exact signal value, but with our 8-bit signed example will always be less than 1 / 128, or 0.0087125. This fluctuating error signal will be seen as a form of noise (called quantization noise), or unwanted signal in the digital video

# Table 11.1

| Signal Level | Closest 8-Bit Representation | Hexadecimal Value | Actual Error | Error As A Percent Of Signal Level |
|---|---|---|---|---|
| 0.50300 | 0.5000000 | 0 × 40 | 0.00300 | 0.596% |
| 0.50300 | 0.5078125 | 0 × 41 | −0.0048128 | 0.957% |
| 0.05030 | 0.0468750 | 0 × 06 | 0.003425 | 6.809% |
| 0.05030 | 0.0546875 | 0 × 07 | −0.0043875 | 8.722% |
| 0.00503 | 0.000000 | 0 × 00 | 0.00503 | 100% |
| 0.00503 | 0.0078125 | 0 × 01 | −0.0027825 | 55.32% |

processing. It can be modeled as an injection of noise when simulated in an algorithm with unlimited numerical precision.

When the signal level is fairly large for the allowable range, (0.503 is close to one half the maximum value), the percentage error is small — less than 1%. As the signal level gets smaller, the error percentage gets larger, as the table indicates.

Quantization noise is always present and is, on average, the same level (any noise-like signal will rise and fall randomly, so we usually concern ourselves with the average level). But as the input signal decreases in level, the quantization noise becomes relatively more significant. Eventually, for very small input signal levels, the quantization noise can become so significant that it degrades the quality of whatever signal processing is to be performed. Think of it as like static on a car radio: as you get further from the radio station, the radio signal gets weaker, and eventually the static noise makes it difficult or unpleasant to listen to, even if you increase the volume.

So what can we do if our signal is sometimes strong (0.503 for example), and sometimes weak (0.00503 for example)? Another way of describing this is to say that the signal has a large dynamic range. The dynamic range describes the ratio between the largest and smallest value of the signal, in this case 100.

Suppose we exchange our 8-bit representation with 12-bit representation? Then our maximum range is still from −1 to +1, but our step size is now 1 / 2048, which is 0.000488. Let's make a 12-bit table similar to the 8-bit example.

This is a significant difference. The actual error is always less than our step size, 1 / 2048. But the error as a percentage of signal level is dramatically improved and this is our concern in signal processing. Because of the much smaller step size of the 12-bit

# Table 11.2

| Signal Level | Closest 12 Bit Representation | Hexadecimal Value | Actual Error | Error As A Percent Of Signal Level |
|---|---|---|---|---|
| 0.50300 | 0.502930 | 0 × 406 | 0.000070 | 0.0139% |
| 0.50300 | 0.503418 | 0 × 407 | −0.000418 | 0.0831% |
| 0.05030 | 0.050293 | 0 × 067 | 0.000007 | 0.0140% |
| 0.05030 | 0.050781 | 0 × 068 | −0.000481 | 0.9568% |
| 0.00503 | 0.004883 | 0 × 00A | 0.000147 | 2.922% |
| 0.00503 | 0.005371 | 0 × 00B | −0.000341 | 6.779% |

representation, the quantization noise is much less, allowing even small signals to be represented with acceptable precision. Another way of describing this is to introduce the concept of signal to noise power ratio, or SNR. This describes the power of the largest signal compared to the background noise. This can be very easily seen on a frequency domain or spectral plot of a signal. There can be many sources of noise, but, for now, we are only considering the quantization noise introduced by the digital representation of the video pixel values.

What we have just described is the uniform quantizer, where all the step sizes are equal. However, there are alternate quantizing mappings, where the step size varies across the signal amplitude. For example, a quantizer could be designed to give the same SNR across the signal range, requiring a small step size when the signal amplitude is small, and a larger step size as the signal increases in value. The idea is to provide a near constant quantization error as a percentage of the signal value. This type of quantizing is performed on the voice signals in the US telephone system, known as μ-law encoding.

Another possible quantization scheme could be to use a small step size for regions of the signal where there is a high probability of signal amplitude occurring, and a larger step for regions where signal amplitude has less likelihood of occurring.

Uniform quantizing is by far the most commonly used scheme in video signal processing, because we are often not encoding simply amplitude, representing small or large signals. In many cases, we could be encoding color information, where the values map to various color intensities, rather than signal amplitude.

If we try to use the likelihood of different values to optimize the quantizer, we have to assume that the probability distribution of the video data is known. Alternatively, the uniform quantizer can be followed by some type of Huffman coding or differential encoding, which will optimize the signal representation for the minimum average number of bits.

Vector Quantization is also commonly used. In the preceding discussion, a single set of values or signals is being quantized with a given bit representation. However, a collection of related signals can be quantized to a single representation using a given number of bits.

For example, if we assume the pixel is represented in RGB format, with 8 bits used for each color, the color red can be represented in $2^8$ or 256 different intensities.

Each pixel uses a total of 24 bits, for a total of $2^{24}$, or about 16 million possible values. Intuitively, this seems excessive as it is unlikely that the human eye can distinguish between that many colors. So a vector quantizer might map the 16 million possible values into a color table of 256 total colors, allowing 256 combinations of red, green and blue combinations. This mapping results in just eight bits to present each pixel.

This seems reasonable, but the complexity lies in mapping 16 million possible inputs to the allowed 256 representations, or color codewords. If done using a look-up table, memory of 16 million bytes would be required for this quantization, with each memory location containing one of the 256 color codewords. This is excessive, so some sort of mapping algorithm or computation is required to map the 16 million possible color combinations to the closest color codeword. It is hopefully becoming clear that most methods to compress video, or other data for that matter, come at the expense of increased complexity and increased computational rates.

## 11.8 Decibels

Signal to Noise Power Ratio (SNR) is usually expressed in decibels (denoted dB), using a logarithmic scale. The SNR of a digitally represented signal can be determined by the following equation:

$$\text{SNR}_{\text{quantization}}\,(\text{dB}) = 6.02 \times (\text{Number of bits}) + 1.76$$

Each additional bit of the signal representation gains 6 dB of SNR. 8-bit representation is capable of a signal with an SNR of about 48 dB, 12-bit can do better at 72 dB, and 16-bit will yield up to 96 dB. This only accounts for the effect of quantization noise: in practice there are other effects which could also degrade SNR in a system.

There is one last important point on decibels. This is a very commonly used term in many areas of digital signal processing subsystems. A decibel is simply a ratio, commonly of amplitude, power or voltage, similar to percentage. But because of the extremely high ratios commonly used (a billion is not uncommon), it is convenient to express this logarithmically. The logarithmic expression also allow chains of circuits, or signal processing operations, each with its own ratio (say of output power to input power), to simply be added up to find the final ratio.

A common area of confusion is differentiating between signal levels or amplitude (voltage if an analog circuit) and signal power. Power measurements are virtual in the digital world, but can be directly measured in analog circuits with which video systems interface, such as the RF amplifiers and analog signal levels for circuits in head-end cable systems, or video satellite transmission.

There are two definitions of dB commonly used:

$dB_{voltage} = dB_{digital\ value} = 20 \times \log$ (voltage signal 1 / voltage signal 2)

$dB_{power} = 10 \times \log$ (power signal 1 / power signal 2)

The designation of "signal 1" and "signal 2" depends on the situation. For example, with an RF power amplifier (analog), the dB of gain will be the 10 log (output power / input power). For digital applications, the dB of SNR will be the 20 log (maximum input signal / quantization noise signal level).

dB can refer to many different ratios in video system designs and it is easy to get confused about whether to use a multiplicative factor of 10 or 20, if the reasoning behind these numbers is unclear.

Voltage squared is proportional to power. If a given voltage is doubled in a circuit, it requires four times as much power. This goes back to a basic Ohm's law equation:

$Power = Voltage^2 / Resistance$

In many analog circuits, signal power is used because that is what the lab instruments work with, and while different systems may use different resistance levels, power is universal (however, 50 ohm is the most common standard in most analog systems).

When voltage is squared, this needs to be taken into account in the computation of logarithmic decibel relation. Remember, log $x^y = y \log x$. Hence, the multiply factor of "2" is required for voltage ratios, changing the "10" to a "20".

In the digital world, the concept of resistance and power do not exist. A given signal has specific amplitude, expressed in a digital numerical system (such as signed fractional or integer for example).

Understanding that dB increases using the two measurement methods is important. Let's look at doubling of the amplitude ratio and doubling of the power ratio:

$6.02 \text{ dB}_{voltage} = 6.02 \text{ dB}_{digital\ value} = 20 \times \log (2 / 1)$
$3.01 \text{ dB}_{power} = 10 \times \log (2 / 1)$

This is why shifting a digital signal left one bit (multiplying by 2) will cause a 6 dB signal power increase, and why so often the term 6 dB/bit is used in conjunction with ADCs, DACs or digital systems in general.

By the same reasoning, doubling in power to an RF engineer means a 3 dB increase. This will impact upon the entire system: coding gain, as used with error correcting code methods, is based upon power. All signals at antenna interfaces are defined in terms of power, and the decibels used will be power ratios.

In both systems, ratios of equal power or voltage are 0 dB. For example, a unity gain amplifier has a gain of 0 dB.

$0 \text{ dB}_{power} = 10 \times \log (1 / 1)$

A loss would be expressed as a negative dB. For example a circuit whose output is equal to ½ the input power:

$-3.01 \text{ dB}_{power} = 10 \times \log (1 / 2)$

# FREQUENCY DOMAIN REPRESENTATION

## CHAPTER OUTLINE

**12.1 DFT and IDFT Equations  85**
    12.1.1 First DFT Example  86
    12.1.2 Second DFT Example  88
    12.1.3 Third DFT Example  89
    12.1.4 Fourth DFT Example  90
**12.2 Fast Fourier Transform  93**
**12.3 Discrete Cosine Transform  97**

In this chapter we are going to examine the DFT (Discrete Fourier Transform) and its more popular cousin, the FFT (Fast Fourier Transform). The FFT is used in many applications, and we will see its uses in broadcast modulation and distribution systems. It also lays the basis for the DCT (Discrete Cosine Transform), commonly used in image compression, which is also covered.

This chapter has more mathematics than most of the others in this book. It requires some familiarity with basic trigonometry, complex numbers and the complex exponential. A refresher on these topics is provided as an appendix for those who haven't used these in a long time.

The DFT is simply a transform. It takes a sequence of sampled data (a signal), and computes the frequency content of that sampled data sequence. This will give the representation of the signal in the frequency domain, as opposed to the familiar time domain representation. This can be done in both the vertical and horizontal dimensions, although for now we will assume a one-dimensional signal. The result is a frequency-domain representation of the signal, providing the spectral content of a given signal. All of the signal information is preserved, but in a different form.

The IDFT (Inverse Discrete Fourier Transform) computes the time-domain representation of the signal from the frequency-domain information. Using these transforms, it is possible to

**Digital Video Processing for Engineers.** http://dx.doi.org/10.1016/B978-0-12-415760-6.00012-X

switch between the time domain signal and the frequency domain spectral representation. The DFT takes a complex signal and decomposes it into a sum of different frequency cosine and sine waves.

In order to appreciate what is happening, we are going to examine a few simple examples. This will involve multiplying and summing up complex numbers, which, while not difficult, can be tedious. We will minimize the tedium by using a short transform length, but it cannot really be avoided in order to understand the DFT.

We will start with the definition of the Fourier transform. The basic Fourier theory tells us that any signal can be represented at a sum of various frequency components, with a particular gain and phase for each component. The signal, f(x), is multiplied and integrated by each possible frequency. Each frequency is represented by the complex exponential, which is just a complex sinusoid. This is also known as the Euler equation:

$$e^{j\omega} = \cos(\omega) + j\sin(\omega)$$

The Fourier transform is defined as:

$$F(\omega) = \int\limits_{-\infty}^{+\infty} f(x)e^{j\omega dx}dx$$

Let's see if we can start simplifying: we will decide to perform the calculation over a finite length of sampled data signal "x", which contains N samples, rather than continuous signal f(x) $C_i$. This gets rid of the infinity, and makes this something we can actually build.

$$F(\omega) = \sum_{i=0}^{N-1} X_i e^{-j\omega i.}$$

Note that $\omega$ is a continuous variable, which we evaluated over a $2\pi$ interval, usually from $-\pi$ to $\pi$. Now we will be transforming a sampled time-domain signal to a sampled frequency-domain spectral plot. So rather than computing $\omega$ continuously from $-\pi$ to $\pi$, we will instead compute $\omega$ at M equally spaced points over an interval of $2\pi$. To avoid aliasing in the frequency domain, we must make $M \geq N$. The reverse transform is the IDFT (or IFFT), which reconstructs the sampled time-domain signal of length N and does not require more than N points in the frequency domain. Therefore we will set $N = M$, and the frequency domain representation will have the same number of points as the time domain representation. We will use the convention "$x_i$" to represent the time-domain version of the signal, and "$X_k$" to represent the frequency-domain representation of the signal. Both indexes i and k will run from 0 to $N-1$.

# 12.1 DFT and IDFT Equations

DFT (time $\rightarrow$ frequency)

$$X_k = H(2\pi k/N) = \sum_{i=0}^{N-1} x_k e^{-j2\pi ki/N} \quad \text{for } k = \{0..N-1\}$$

IDFT (frequency $\rightarrow$ time)

$$X_i = 1/N \times \sum_{k=0}^{N-1} X_k e^{+j2\pi ki/N} \quad \text{for } i = \{0..N-1\}$$

These equations do appear very similar. The differences are the negative sign on the exponent on the DFT equation, and the factor of 1 / N on the IDFT equation. The DFT equation requires that every single sample in the frequency domain has a contribution from each and every one of the time domain samples. And the IDFT equation requires that every single sample in the time domain has a contribution from each and every one of the frequency domain samples. To compute a single sample of either transform requires N complex multiplies and additions. To compute the entire transform will require computing N samples, for a total of $N^2$ multiplies and additions. This can become a computational problem when N becomes large. As we will see later, this is the reason the FFT and IFFT were developed.

The values of $X_k$ represent the amount of signal energy at each frequency point. Imagine taking a spectrum of 1 MHz which we divide into N bins. If $N = 20$, then we will have 20 frequency bins, each 50 kHz wide. The DFT output, $X_k$, will represent the signal energy in each of these bins. For example, $X_0$ will represent the signal energy at 0 kHz, or the DC component of the signal. $X_1$ will represent the frequency content of the signal at 50 kHz. $X_2$ will represent the frequency content of the signal at 100 kHz. $X_{19}$ will represent the frequency content of the signal at 950 kHz.

A few comments on these transforms might prove helpful. Firstly, they are reversible. We can take a signal represented by N samples, and perform the DFT on it. We will get N outputs representing the frequency response or spectrum on the signal. If we take this frequency response and perform the IDFT on it, we will get back our original signal of N samples back. Secondly, when the DFT output gives the frequency content of the input signal, it is assuming that the input signal is periodic in N. To put it another way, the frequency response is actually the frequency response of an infinite long periodic signal, where the N long sequence of $x_i$ samples repeat over and over. Lastly, the input signal $x_i$ is usually assumed to be a complex (real and quadrature) signal. The frequency response samples $X_k$ are also complex. Often we are more interested in only the magnitude of the frequency response $X_k$, which can be more easily displayed. But in order to get back the original complex input $x_i$ using the IDFT, we would need the complex sequence $X_k$.

At this point, we will do a few examples, selecting $N = 8$.

For our $N = 8$ point DFT, the output gives us the distribution of input signal energy into eight frequency bins, corresponding to the frequencies in Table 12.1 below. By computing the DFT coefficients $X_k$, we are performing a correlation, or trying to match, our input signal to each of these frequencies. The magnitude DFT output coefficients $X_k$ represent the degree of match of the time-domain signal $x_i$ to each frequency component.

## 12.1.1 First DFT Example

Let's start with a simple time domain signal consisting of {1,1,1,1,1,1,1,1}. Remember, the DFT assumes this signal keeps repeating, so the frequency output will actually be that of an indefinite string of 1s. As this signal is unchanging, then by intuition, we will expect that zero frequency component (DC of

# Table 12.1

| k | $X_k$ | Compute By Correlating To Complex Exponential Signal | △ Phase Between Each Sample Of Complex Exponential Signal |
|---|---|---|---|
| 0 | $X_0$ | $e^0$ for $i = 0,1..7$ | 0 |
| 1 | $X_1$ | $e^{-j2\pi i/8}$ for $i = 0,1..7$ | $-\pi/4$ or $-45$ degrees |
| 2 | $X_2$ | $e^{-j4\pi i/8}$ for $i = 0,1..7$ | $-2\pi/4$ or $-90$ degrees |
| 3 | $X_3$ | $e^{-j6\pi i/8}$ for $i = 0,1..7$ | $-3\pi/4$ or $-135$ degrees |
| 4 | $X_4$ | $e^{-j8\pi i/8}$ for $i = 0,1..7$ | $-4\pi/4$ or $-180$ degrees |
| 5 | $X_5$ | $e^{-j10\pi i/8}$ for $i = 0,1..7$ | $-5\pi/4$ or $-225$ degrees |
| 6 | $X_6$ | $e^{-j12\pi i/8}$ for $i = 0,1..7$ | $-6\pi/4$ or $-270$ degrees |
| 7 | $X_7$ | $e^{-j14\pi i/8}$ for $i = 0,1..7$ | $-7\pi/4$ or $-315$ degrees |

signal) is going to be the only non-zero component of the DFT output $X_k$.

Starting with $X_k = \sum_{i=0}^{N-1} x_i e^{-j2\pi ki/N}$ and setting $N = 8$ and all $x_i = 1$:

$X_k = \sum_{i=0}^{7} 1 \times e^{-j2\pi ki/8}$, and setting $k=0$ (recall that $e^0=1$)

$$X_0 = \sum_{i=0}^{7} 1 \times 1 = 8$$

Next, evaluate for $k = 1$:

$$X_1 = \sum_{i=0}^{7} 1 \times e^{-j2\pi i/8}$$
$$= 1 + e^{-j2\pi/8} + e^{-j4\pi/8} + e^{-j6\pi/8} + e^{-j8\pi/8} + e^{-j10\pi/8}$$
$$+ e^{-j12\pi/8} + e^{-j14\pi/8}$$

$$X_1 = 1 + (0.7071 - j0.7071) - j + (-0.7071 - j0.7071) - 1$$
$$+ (-0.7071 + j0.7071) + j + (0.7071 + j0.7071)$$

$X_1 = 0$

The eight terms of the summation for $X_1$ cancel out. This makes sense because it's a sum of eight equally spaced points about the origin on the unit circle of the complex plane. The summation of these points must equal the center, in this case zero.

Next, evaluate for $k = 2$:

$$X_2 = \sum_{i=0}^{7} 1 \times e^{-j2\pi i/8}$$
$$= 1 + e^{-j\pi/2} + e^{-j\pi} + e^{-j3\pi/2} + e^{-j2\pi} + e^{-j5\pi/2} + e^{-j3\pi}$$
$$+ e^{-j7\pi/2}$$

$$X_2 = 1 - j - 1 + j + 1 - j - 1 + j = 0$$

We will find out similarly that $X_3$, $X_4$, $X_5$, $X_6$, $X_7$ are also zero. Each of these will represent eight points equally spaced about the unit circle. $X_1$ has its points spaced at $-45°$ increments, $X_2$ has its points spaced at $-90°$ increments, $X_3$ has its points spaced at $-135°$ increments, and so forth (the points may wrap multiple times around the unit circle in the frequency domain). So, as we expected, the only non-zero term is $X_0$, which is the DC term. There is no other frequency content of the signal.

Now, let us use the IDFT to get the original sequence back:

$$x_i = 1/N \times \sum_{k=0}^{N-1} X_k e^{+j2\pi ki/N} \text{ for } N = 8 \text{ and } X_0$$
$$= 8, \text{ all other } X_k = 0$$

$$x_i = 1/8 \times \sum_{k=0}^{N-1} X_k \, e^{+j2\pi ki/8}$$

Since $X_0 = 8$ and the rest are zero, we only need to evaluate the summation for $k = 0$:

$$x_i = 1/8 \times 8 \times e^{+j2\pi 0i/8} = 1$$

This is true for all values of i (the 0 in the exponent means the value of "i" is irrelevant). So we get an infinite sequence of 1s.

In general however, we would evaluate for i from 0 to $N-1$. Due to the periodicity of the transform, there is no point in evaluating when $i = N$ or greater. If we evaluate for $i = N$, we will find we get the same value as $i = 0$, and for $i = N + 1$, we will get the same value as $i = 1$.

## 12.1.2 Second DFT Example

Let's consider another simple example, with a time domain signal $\{1, j, -1, -j, 1, j, -1, -j\}$. This is actually the complex exponential $e^{+j2\pi i/4}$. This signal consists of a single frequency, and corresponds to one of the frequency "bins" that the DFT will measure. So we can expect a non-zero DFT output in this frequency bin, but zero elsewhere. Let's see how this works out.

Starting with $X_k = \sum_{i=0}^{N-1} x_i \, e^{-j2\pi ki/N}$ and setting $N = 8$ and $x_i = \{1, j, -1, -j, 1, j, -1, -j\}$:

$$X_0 = \sum_{i=0}^{7} x_i \times 1 \, , \text{ as } k = 0 (e^0 = 1)$$

$X_0 = 1 + j - 1 - j + 1 + j - 1 - j = 0$, so the signal has no DC content, as expected. Notice that to calculate $X_0$ — which is the DC content of $x_I$ — the DFT just sums (essentially averaging) the input samples.

Next, evaluate for $k = 1$:

$$X_1 = \sum_{i=0}^{7} x_i \times e^{-j2\pi i/8}$$
$$= 1 \times 1 + j \times e^{-j2\pi/8} - 1 \times e^{-j4\pi/8} - j \times e^{-j6\pi/8} + 1 \times e^{-j8\pi/8}$$
$$+ j \times e^{-j10\pi/8} - 1 \times e^{-j12\pi/8} - j \times e^{-j14\pi/8}$$

$$X_1 = 1 + [j \times (0.7071 - j0.7071)] + j - [j \times (-0.7071 - j0.7071)]$$
$$- 1 + [j \times (-0.7071 + j0.7071)] - j - [j \times (0.7071 + j0.7071)]$$

$$X_1 = 0$$

If you take the time to work this out, you will see that all eight terms of the summation cancel out. This also happens for $X_3$, $X_4$,

$X_5$, $X_6$, and $X_7$. Let's look at $X_2$ now. We will also express $x_i$ using complex exponential format of $e^{+j2\pi i/4}$.

$$X_k = \sum_{i=0}^{7} x_i \, e^{-j2\pi ki/8}$$

$$X_2 = \sum_{i=0}^{7} x_i \times e^{-j4\pi i/8} = \sum_{i=0}^{7} e^{+j2\pi i/4} \times e^{-j4\pi i/8}$$

$$= \sum_{i=0}^{7} e^{+j2\pi i/4} \times e^{-j2\pi i/4}$$

Remember that when exponentials are multiplied the exponents are added ($x^2 \times x^3 = x^5$). Here the exponents are identical, except of opposite sign, so they add to zero.

$$X_2 = \sum_{i=0}^{7} e^{+j2\pi i/4} \times e^{-j2\pi i/4} = \sum_{i=0}^{7} e^0 = \sum_{i=0}^{7} 1 = 8$$

The sole frequency component of the input signal is $X_2$ because our input is a complex exponential frequency at the exact frequency that $X_2$ represents.

## 12.1.3 Third DFT Example

Next, we can try modifying $x_i$ such that we introduce a phase shift, or delay (like substituting a sine wave for a cosine wave). If we introduce a delay, so $x_i$ starts at j instead of 1, but is still the same frequency, the input $x_i$ is still rotating around the complex plane at the same rate, but starts at j (angle of $\pi / 2$) rather than 1 (angle of 0). Now the sequence $x_i = \{j,-1,-j,1,j,-1,-j,1\}$ or $e^{+j(2\pi(i+1)/4)}$.

The DFT output will result in $X_0$, $X_1$, $X_3$, $X_4$, $X_5$, $X_6$, and $X_7 = 0$, as before. Changing the phase cannot cause any new frequency to appear in the other bins.

Next, evaluate for $k = 2$:

$$X_k = \sum_{i=0}^{7} x_i \, e^{-j2\pi ki/8}$$

$$X_2 = \sum_{i=0}^{7} x_i \times e^{-j4\pi i/8} = \sum_{i=0}^{7} e^{+j(2\pi(i+1)/4)+1)} \times e^{-j4\pi i/8}$$

We need to sum the two values of the two exponents:

$$+j(2\pi(i+1)/4) + -j4\pi i/8 = +j2\pi i/4 + j2\pi/4 - j2\pi i/4 = j\pi/2$$

Substituting back this exponent value:

$$X_2 = \sum_{i=0}^{7} e^{+j((2\pi i/4)+1)} \times e^{-j4\pi i/8} = \sum_{i=0}^{7} e^{+j\pi/2} = \sum_{i=0}^{7} j = j8$$

So we get exactly the same magnitude at the frequency component $X_2$. The difference is the phase of $X_2$. Therefore we can see that the DFT does not just pick out the frequency components of a signal, but is sensitive to the phase of those components. The phase, as well as amplitude of the frequency components $X_k$, can be represented because the DFT output is complex.

The process of the DFT is to correlate the N sample input data stream $x_i$ against N equally spaced complex frequencies. If the input data stream is one of these N complex frequencies, then we will get a perfect match, and get zero in the other N−1 frequencies which don't match. But what happens if we have an input data stream with a frequency in between one of the N frequencies?

To review, we have looked at three simple examples. The first was a constant level signal, so the DFT output was just the zero frequency or DC component. The second example was a complex frequency which matched exactly to one of the frequency bins, $X_k$, of the DFT. The third was the same complex frequency, but with a phase offset. The fourth will be a complex frequency not matched to one of the N frequencies used by the DFT.

### 12.1.4 Fourth DFT Example

We will look at an input signal of frequency $e^{+j2.1\pi i / 8}$. This is rather close to $e^{+j2\pi i / 8}$, so we would expect a rather strong output at $X_1$. Let's see what the N = 8 DFT result is — hopefully the arithmetic is all correct. Slogging through this arithmetic is purely optional — the details are shown to provide a complete example.

Generic DFT equation for N = 8: $X_k = \sum_{i=0}^{7} x_i e^{-j2\pi ki/8}$

$$X_0 = \sum_{i=0}^{7} e^{+j2.1\pi i/8} \times 1 = \sum_{i=0}^{7} e^{+j2.1\pi i/8}$$

$$= [1 + j0] + [0.6788 + j0.7343] + [-0.0785 + j0.9969]$$
$$+ [-0.7853 + j0.6191] + [-0.9877 - j0.1564] + [-0.5556$$
$$- j0.8315] + [0.2334 - j0.9724] + [0.8725 - j0.4886]$$
$$= 0.3777 - j0.0986$$

$$X_1 = \sum_{i=0}^{7} e^{+j2.1\pi i/8} \times e^{-j2\pi i/8}$$

$$= \sum_{i=0}^{7} e^{+j0.1\pi i/8} = [1 + j0] + [0.9992 + j0.0393] + [0.9969$$
$$+ j0.0785] + [0.9931 + j0.1175] + [0.9877 + j0.1564]$$
$$+ [0.9808 + j0.1951] + [0.9724 + j0.2334] + [0.9625$$
$$+ j0.2714]$$
$$= 7.8925 + j1.0917$$

$$X_2 = \sum_{i=0}^{7} e^{+j2.1\pi i/8} \times e^{-j4\pi i/8} = \sum_{i=0}^{7} e^{-j1.9\pi i/8}$$

$$= [1+j0] + [0.7343 - j0.6788] + [0.0785 - j0.9969] + [-0.6191$$

$$- j0.7853] + [-0.9877 - j0.1564] + [-0.8315 + j0.5556]$$

$$+ [-0.2334 + j0.9724] + [0.4886 + j0.8725]$$

$$= -0.3703 - j0.2170$$

$$X_3 = \sum_{i=0}^{7} e^{+j2.1\pi i/8} \times e^{-j6\pi i/8} = \sum_{i=0}^{7} e^{-j3.9\pi i/8}$$

$$= [1+j0] + [0.0393 - j0.9992] + [-0.9969 - j0.0785]$$

$$+ [-0.1175 + j0.9931] + [0.9877 + j0.1564] + [0.1951$$

$$- j0.9808] + [-0.9724 - j0.2334] + [-0.2714 + j0.9625]$$

$$= -0.1362 - j0.1800$$

$$X_4 = \sum_{i=0}^{7} e^{+j2.1\pi i/8} \times e^{-j8\pi i/8} = \sum_{i=0}^{7} e^{-j5.9\pi i/8}$$

$$= [1+j0] + [-0.6788 - j0.7343] + [-0.0785 + j0.9969]$$

$$+ [0.7853 - j0.6191] + [-0.9877 - j0.1564] + [0.5556$$

$$+ j0.8315] + [0.2334 - j0.9724] + [-0.8725 + j0.4886]$$

$$= -0.0431 - j0.1652$$

$$X_5 = \sum_{i=0}^{7} e^{+j2.1\pi i/8} e^{-j10\pi i/8} = \sum_{i=0}^{7} e^{-j7.9\pi i/8}$$

$$= [1+j0] + [-0.9992 - j0.0393] + [0.9969 + j0.0785]$$

$$+ [-0.9931 - j0.1175] + [0.9877 + j0.1564] + [-0.9808$$

$$- j0.1951] + [0.9724 + j0.2334] + [-0.9625 - j0.2714]$$

$$= 0.0214 - j0.1550$$

$$X_6 = \sum_{i=0}^{7} e^{+j2.1\pi i/8} \times e^{-j12\pi i/8} = \sum_{i=0}^{7} e^{-j9.9\pi i/8}$$

$$= [1+j0] + [-0.7343 + j0.6788] + [0.0785 - j0.9969] + [0.6191$$

$$+ j0.7853]$$

$$= 0.0849 + -j0.1449$$

$$X_7 = \sum_{i=0}^{7} e^{+j2.1\pi i/8} \times e^{-j14\pi i/8} = \sum_{i=0}^{7} e^{-j11.9\pi i/8}$$

$$= [1+j0] + [-0.0393 + j0.9992] + [-0.9969 - j0.0785]$$

$$+ [0.1175 - j0.9931] + [0.9877 + j0.1564] + [-0.1951$$

$$+ j0.9808] + [-0.9724 - j0.2334] + [0.2714 - j0.9625]$$

$$= 0.1730 - j0.1310$$

That was a bit tedious, but there is some insight to be gained from the results of these simple examples.

Table 12.2 shows how the DFT is able to represent the signal energy in each frequency bin. The first example has all the energy at DC. The second and third examples are complex exponentials at frequency $\omega = \pi / 2$ radians/sample, which corresponds to DFT output $X_2$. The magnitude of the DFT outputs is the same for both examples, since the only difference of the inputs is the phase. The fourth example is the most interesting. In this case, the input frequency is close to $\pi / 4$ radians/sample, which corresponds to DFT output $X_1$. So $X_1$ does capture most of the energy of the

## Table 12.2

| DFT Output Magnitude | $x_i = \{1,1,1,1,1,1,1,1\}$ | $x_i = e^{+j2\pi i/4}$ | $x_i = e^{+j(2\pi(i+1)/4)}$ | $x_i = e^{+j2.1\pi i/8}$ |
|---|---|---|---|---|
| Output $X_0$ | 8 | 0 | 0 | 0.39 |
| Output $X_1$ | 0 | 0 | 0 | 7.99 |
| Output $X_2$ | 0 | 8 | 8 | 0.43 |
| Output $X_3$ | 0 | 0 | 0 | 0.23 |
| Output $X_4$ | 0 | 0 | 0 | 0.17 |
| Output $X_5$ | 0 | 0 | 0 | 0.16 |
| Output $X_6$ | 0 | 0 | 0 | 0.17 |
| Output $X_7$ | 0 | 0 | 0 | 0.22 |

signal. But small amounts of energy spill into other frequency bins, particularly the adjacent bins.

We can increase the frequency sorting ability of the DFT by increasing the value of N. This will narrow each frequency bin because the frequency spectrum is divided into N sections in the DFT. This will result in a given frequency component being more selectively represented by a particular frequency bin. For example, the frequency response plots of the filters contained in the FIR chapter are computed with a value of N equal to 1024. This means the spectrum was divided into 1024 sections, and the response computed for each particular frequency. When plotted together, this gives a very good representation of the complete frequency spectrum.

Please note this also requires taking a longer input sample stream $x_i$, equal to N. This, in turn, will require a much greater number of operations to compute.

## 12.2 Fast Fourier Transform

At some point, some smart people searched for a way to compute the DFT in a more efficient way. The result is the FFT, or Fast Fourier Transform. Rather than requiring $N^2$ complex multiplies and additions, the FFT requires $N \times \log_2 N$ complex multiplies and additions. This may not sound like a big deal, but look at the comparison in Table 12.3 below.

So by using the FFT algorithm on a 1024 point (or sample) input, we are able to reduce the computational requirements to less than 1%, or by two orders of magnitude.

The FFT achieves this by reusing previously calculated interim products. We can see this through a simple example.

Let's start with the calculation of the simplest DFT: N = 2 DFT. Generic DFT equation for N = 2 : $X_k = \sum_{i=0}^{1} x_i \ e^{-j2\pi ki/2}$

$$X_0 = x_0 \times e^{-j2\pi 0/2} + x_1 \times e^{-j2\pi 0/2}$$

$$X_1 = x_0 \times e^{-j2\pi 0/2} + x_1 \times e^{-j2\pi 1/2}$$

As $e^0 = 1$, we can simplify to find:

$$X_0 = x_0 + x_1$$

$$X_1 = x_0 + x_1 \times e^{-j\pi} = x_0 - x_1$$

Next, we will do the four point (N = 4) DFT. Generic DFT equation for N = 4 : $X_k = \sum_{i=0}^{3} x_i \ e^{-j2\pi ki/4}$

## Table 12.3

| N | Dft — $N^2$ Complex Multiplies & Additions | Fft — $N \times \log_2 n$ Complex Multiplies & Additions | Computational Effort Of Fft Compared To Dft |
|---|---|---|---|
| 8 | 64 | 24 | 37.50% |
| 32 | 1024 | 160 | 15.62% |
| 256 | 65,536 | 2048 | 3.12% |
| 1024 | 1,048,576 | 10,240 | 0.98% |
| 4096 | 16,777,216 | 49,152 | 0.29% |

$$X_0 = x_0 \times e^{-j2\pi0/4} + x_1 \times e^{-j2\pi0/4} + x_2 \times e^{-j2\pi0/4} + x_3 \times e^{-j2\pi0/4}$$

$$X_1 = x_0 \times e^{-j2\pi0/4} + x_1 \times e^{-j2\pi1/4} + x_2 \times e^{-j2\pi2/4} + x_3 \times e^{-j2\pi3/4}$$

$$X_2 = x_0 \times e^{-j2\pi0/4} + x_1 \times e^{-j2\pi2/4} + x_2 \times e^{-j2\pi4/4} + x_3 \times e^{-j2\pi6/4}$$

$$X_3 = x_0 \times e^{-j2\pi0/4} + x_1 \times e^{-j2\pi3/4} + x_2 \times e^{-j2\pi6/4} + x_3 \times e^{-j+\pi9/4}$$

The term $e^{-j2\pi k/4}$ repeats itself with a period of $k = 4$, as the complex exponential makes a complete circle and begins another. This periodicity means that $e^{-j2\pi k/4}$ is equal when evaluated for $k = 0, 4, 8, 12...$ It is again equal for $k = 1, 5, 9, 13...$ Because of this, we can simplify the last two terms of expressions for $X_2$ and $X_3$ (shown in bold below). We can also remove the exponential when it is to the power of zero.

$$X_0 = x_0 + x_1 + x_2 + x_3$$

$$X_1 = x_0 + x_1 \times e^{-j2\pi1/4} + x_2 \times e^{-j2\pi2/4} + x_3 \times e^{-j2\pi3/4}$$

$$X_2 = x_0 + x_1 \times e^{-j2\pi2/4} + x_2 \times e^{-j2\pi0/4} + x_3 \times e^{-j2\pi2/4}$$

$$X_3 = x_0 + x_1 \times e^{-j2\pi3\{ts\}/4} + x_2 \times e^{-j2\pi2/4} + x_3 \times e^{-j2\pi1/4}$$

Now we are going to rearrange the terms of the four point ($N = 4$) DFT, grouping the even and odd terms together.

$$X_0 = [x_0 + x_2] + [x_1 + x_3]$$

$$X_1 = [x_0 + x_2 \times e^{-j2\pi2/4}] + [x_1 \times e^{-j2\pi1/4} + x_3 \times e^{-j2\pi3/4}]$$

$$X_2 = [x_0 + x_2] + [x_1 \times e^{-j2\pi2/4} + x_3 \times e^{-j2\pi2/4}]$$

$$X_3 = [x_0 + x_2 \times e^{-j2\pi2/4}] + [x_1 \times e^{-j2\pi3/4} + x_3 \times e^{-j2\pi1/4}]$$

Next we will factor $x_1$ and $x_3$ to get this particular form:

$$X_0 = [x_0 + x_2] + [x_1 + x_3]$$

$$X_1 = [x_0 + x_2 \times e^{-j2\pi2/4}] + [x_1 \times e^{-j2\pi1/4} + x_3 \times e^{-j2\pi3/4}]$$
$$= [x_0 + x_2 \times e^{-j2\pi2/4}] + [x_1 + x_3 \times e^{-j2\pi2/4}] \times e^{-j2\pi/4}$$

$$X_2 = [x_0 + x_2] + [x_1 \times e^{-j2\pi2/4} + x_3 \times e^{-j2\pi2/4}]$$
$$= [x_0 + x_2] + [x_1 + x_3] \times e^{-j2\pi2/4}$$

$$X_3 = [x_0 + x_2 \times e^{-j2\pi 2/4}] + [x_1 \times e^{-j2\pi 3/4} + x_3 \times e^{-j2\pi 1/4}]$$
$$= [x_0 + x_2 \times e^{-j2\pi 2/4}] + [x_1 + x_3 \times e^{-j2\pi 2/4}] \times e^{-j2\pi 3/4}$$

Here is the result:

$$X_0 = [x_0 + x_2] + [x_1 + x_3]$$

$$X_1 = [x_0 + x_2 \times e^{-j2\pi 2/4}] + [x_1 + x_3 \times e^{-j2\pi 2/4}] \times e^{-j2\pi/4}$$

$$X_2 = [x_0 + x_2] + [x_1 + x_3] \times e^{-j4\pi/4}$$

$$X_3 = [x_0 + x_2 \times e^{-j2\pi 2/4}] + [x_1 + x_3 \times e^{-j2\pi 2/4}] \times e^{-j6\pi/4}$$

Now comes the insightful part. Comparing the four equations above, you can see that the bracketed terms used for $X_0$ and $X_1$ are also present in $X_2$ and $X_3$. So we don't need to recompute these terms during the calculation of $X_2$ and $X_3$. We can simply multiply them by the additional exponential outside the brackets. This reusing of partial products in multiple calculations is the key to understanding the FFT efficiency, so at the risk of being repetitive, this is shown again more explicitly below.

$\rightarrow$ define $A = [x_0 + x_2]$, $B = [x_1 + x_3]$
$$X_0 = [x_0 + x_2] + [x_1 + x_3] = A + B$$

$\rightarrow$ define $C = [x_0 + x_2 \times e^{-j2\pi 2/4}]$, $D = [x_1 + x_3 \times e^{-j2\pi 2/4}]$
$$X_1 = [x_0 + x_2 \times e^{-j2\pi 2/4}] + [x_1 + x_3 \times e^{-j2\pi 2/4}] \times e^{-j2\pi/4}$$
$$= C + D \times e^{-j2\pi/4}$$

$$X_2 = [x_0 + x_2] + [x_1 + x_3] \times e^{-j4\pi/4} = A + B \times e^{-j4\pi/4}$$

$$X_3 = [x_0 + x_2 \times e^{-j2\pi 2/4}] + [x_1 + x_3 \times e^{-j2\pi 2/4}] \times e^{-j6\pi/4}$$
$$= C + D \times e^{-j6\pi/4}$$

This process quickly gets out of hand for anything larger than a four point ($N = 4$) FFT. So we are going to use a type of representation called a flow graph, shown in Figure 12.1 below.

The flow graph is an equivalent way of representing the equations, and moreover, represents the actual organization of the computations. You can check the simple example above to see the flow graph gives the same results as the DFT equations. For example, $X_0 = x_0 + x_1 + x_2 + x_3$, and by examining the flow

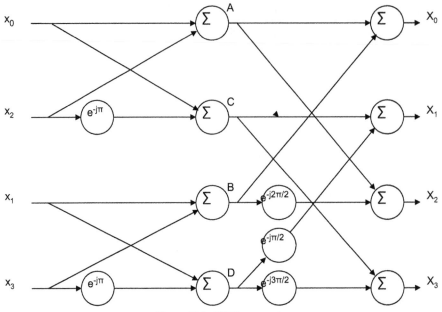

**Figure 12.1.** FFT flow graph.

graph, it is apparent that $X_0 = A + B = [x_0 + x_2] + [x_1 + x_3]$, which is the same result. The order of computations would be to compute pairs {A, C} and {B, D} in the first stage. The next stage would be to compute {$X_0$, $X_2$} and {$X_1$, $X_3$}.

These stages (our example above has two stages) are composed of "butterflies". Each butterfly has two complex inputs and two complex outputs. The butterfly involves one or two complex multiplications and two complex additions. In the first stage, there are two butterflies to compute the two pairs {A, C} and {B, D}. In the second stage, there are two butterflies to compute the two pairs {$X_0$, $X_2$} and {$X_1$, $X_3$}. The complex exponentials multiplying the data path are known as "twiddle factors". In higher N count FFTs, they are simply sine and cosine values that are usually stored in a table.

Now we can see why the FFT is effective in reducing the number of computations. Each time the FFT doubles in size (N increases by a factor of two), we need to add one more stage. Thus, a four-point FFT requires two stages; an eight point FFT requires three stages and a 16 point FFT requires four stages, and so on. The amount of computations required for each stage is proportional to N. The required number of stages is equal to $\log_2 N$. Therefore, the FFT computational load increases by $N \times \log_2 N$. The DFT computational load increases as $N^2$.

This is also the reason why FFT sizes are almost always in powers of 2 (2, 4, 8, 16, 32, 64..). This is sometimes called a "radix 2" FFT. So rather than a 1000 point FFT, one will see a 1024 point FFT. In practice, this common restriction to powers of two is not a problem.

# 12.3 Discrete Cosine Transform

Next, we will discuss the Discrete Cosine Transform, or DCT. Until now we have been talking about the DFT or FFT operating on one-dimensional signals, which are often complex (may have real and quadrature components). The DCT is usually used for two-dimensional signals, such as an image presented by a rectangular array of pixels, and which is a real signal only. When we discuss frequency, it will be in the context of how rapidly the sample values change. We will be sampling spatially across the image in either the vertical or horizontal direction with the DCT.

The DCT is usually applied across an N by N array of pixel data. For example, if we take a region composed of eight by eight pixels, or 64 pixels total, we can transform this into a set of 64 DCT coefficients, which is the spatial frequency representation of the eight by eight region of the image. This is very similar to the DCT. However, instead of expressing the signal as a combination of the complex exponentials of various frequencies, we will be expressing the image data as a combination of cosines of various frequencies, in both vertical and horizontal dimensions.

DFT representation is for a periodic signal, or one that is assumed to be periodic. Imagine connecting a series of identical signals together, end to end. Where the end of the sequence connects to the beginning of the next, there will be a discontinuity, or a step function. This will represent high frequency.

For the DCT, we make an assumption that the signal is folded over on itself. So an eight-long signal depicted becomes 16-long when appended as flipped. This 16-long signal is then symmetric about the midpoint. This is the same property of cosine waves. A cosine is symmetric about the midpoint, which is at $\pi$ (since the period is from 0 to $2\pi$). This property is preserved for higher frequency cosines, as shown by the figures below, showing the sampled cosine waves. The waveform is eight samples long, and if folded over will create a 16-long sampled waveform which will be symmetric, start and end with the same value and has "$u$" cycles across the 16 samples.

To continue requires explanation of some terminology. The value of a pixel at row x and column y is designated as f(x, y), as shown in Figure 12.3. We will compute the DCT coefficients,

$F(u, v)$ using equations which will correlate the pixels to the vertical and horizontal cosine frequencies. In the equations, "$u$" and "$v$" correspond to both the indices in the DCT array, and the cosine frequencies as shown in Figure 12.3.

The relationship is given in the DCT equation, shown below for the eight by eight size.

$$F(u, v) = ¥C_u\,C_v \sum_{x=0}^{7}\sum_{y=0}^{7} f(x, y) \times \cos((2x + 1)u\pi/16)$$
$$\times \cos((2y + 1)v\pi/16)$$

where

$$C_u = \sqrt{2}/2 \text{ when } u = 0, C_u = 1 \quad \text{when } u = 1..7$$

$$C_v = \sqrt{2}/2 \text{ when } v = 0, C_u = 1 \quad \text{when } v = 1..7$$

$f(x,y) =$ the pixel value at that location

This represents 64 different equations, for each combination of $u$, $v$. For example:

$$F(0,0) = 1/8 \times \sum_{x=0}^{7}\sum_{y=0}^{7} f(x, y)$$

Simply put, the summation of all 64 pixel values divided by eight. $F(0, 0)$ is the DC level of the pixel block.

$$F(4,2) = 1/4 \times \sum_{x=0}^{7}\sum_{y=0}^{7} f(x, y) \times \cos((2x + 1)4\pi/16$$
$$\times \cos((2y + 1)2\pi/16)$$

The nested summations indicate that for each of the 64 DCT coefficients, we need to perform 64 summations. This requires $64 \times 64 = 4096$ calculations, which is very process intensive.

The DCT is a reversible transform (provided enough numerical precision is used), and the pixels can be recovered from the DCT coefficients as shown below:

$$f(x, y) = 1/4C_u C_v \sum_{u=0}^{7}\sum_{v=0}^{7} F(u, v) \times \cos((2x + 1)u\pi/16)$$
$$\times \cos((2y + 1)v\pi/16)$$

$$C_u = \sqrt{2}/2 \quad \text{when } u = 0, C_u = 1 \quad \text{when } u = 1..7$$

$$C_v = \sqrt{2}/2 \quad \text{when } v = 0, C_v = 1 \quad \text{when } v = 1..7$$

Another way to look at the DCT is through the concept of basis functions.

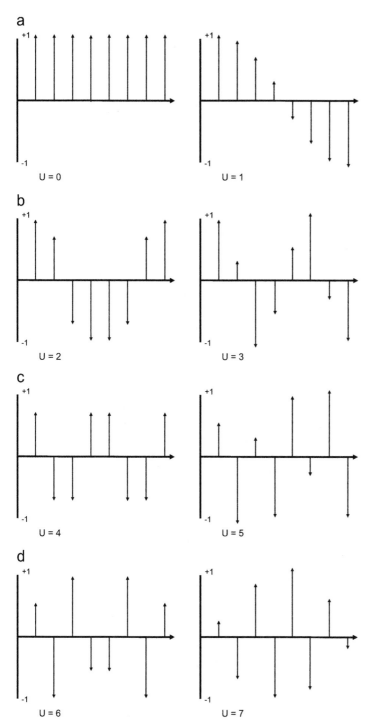

**Figure 12.2a—12.2d.** Sample cosine frequencies used in DCT.

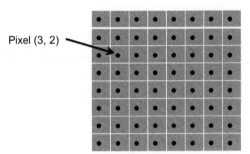

Pixel (3, 2)

**Figure 12.3.** Block of pixels used to compute DCT.

The tiles represent the video pixel block for each DCT coefficient. If $F(0, 0)$ is non-zero, and the rest of the DCT coefficients equal zero, the video will appear as the {0, 0} tile in Figure 12.4. This particular tile is a constant value in all 64 pixel locations, which is what is expected since all the DCT coefficients with some cosine frequency content are zero.

If $F(7, 7)$ is non-zero, and the rest of the DCT coefficients equal zero, the video will appear as the {7, 7} tile in Figure 12.4, which shows high frequency content in both vertical and horizontal direction. The idea is that any block of eight by eight pixels, no matter what the image, can be represented as the weighted sum of these 64 tiles in Figure 12.4.

The DCT coefficients, and the video tiles they represent, form a set of basis functions. From linear algebra, any set of function values f(x,y) can be represented as a linear combination of the basis functions.

The whole purpose of this is to provide an alternate representation of any set of pixels, using the DCT basis functions. By itself, this exchanges one set of 64 pixel values with a set of 64

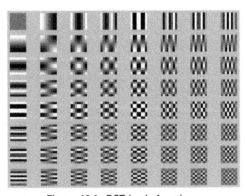

**Figure 12.4.** DCT basis functions.

DCT values. However, in many image blocks many of the DCT values are near zero, or small, and can be presented with few bits. This can allow the pixel block to be presented more efficiently with fewer values. However, this representation is approximate, because when we minimize the bits representing various DCT coefficients, we are quantizing. This is a loss of information, meaning the pixel block cannot be restored perfectly.

# IMAGE COMPRESSION FUNDAMENTALS

## CHAPTER OUTLINE

13.1 Baseline JPEG   103
13.2 DC Scaling   104
13.3 Quantization Tables   104
13.4 Entropy Coding   106
13.5 JPEG Extensions   108

Now that we have the basics of entropy, predictive coding, DCT and quantization, we are ready to discuss image compression. This deals with a single, still image rather than the continuous sequence of images which makes up a video stream.

JPEG is often ubiquitous with image compression. JPEG stands for Joint Photographic Experts Group, a committee that has published international standards on image compression. JPEG is an extensive portfolio of both lossy and lossless image compression standards and options. In this section, we will focus on baseline JPEG.

## 13.1 Baseline JPEG

Baseline JPEG compresses each color plane independently.

A monochrome image would have eight bits per pixel. Generally, lossy compression techniques can represent the image data using less than one bit per pixel, and still give high quality.

RGB images have each of the three color planes treated independently. In YCrCb representation, Y, Cr and Cb are treated independently. For 4:2:2 or 4:2:0 YCrCb, the Cr and Cb are undersampled, and these undersampled color planes will be compressed. For example, standard definition images are 720 (width) by 480 (height) pixels. For 4:2:2 YCrCb representation,

Digital Video Processing for Engineers. http://dx.doi.org/10.1016/B978-0-12-415760-6.00013-1

the Cr and Cb planes will be 360 by 480 pixels. Therefore, a higher degree of compression can be achieved using JPEG on 4:2:2 or 4:2:0 YCrCb images. Intuitively, this makes sense, as more bits are used to represent the luminance to which the human eye is more sensitive, and less for the chrominance to which the human eye is less sensitive.

## 13.2 DC Scaling

Each color plane of the image is divided up into $8 \times 8$ pixel blocks. Each 8-bit pixel can have a value ranging from 0 to 255. The next step is to subtract 128 from all 64 pixel values, so the new range is $-128$ to $+127$. The $8 \times 8$ DCT is next applied to this set of 64 pixels. This DCT output is the frequency domain representation of the image block.

The upper left DCT output is the DC value, or average of all the 64 pixels. Since we subtracted 128 prior to the DCT processing, the DC value can range from $-1024$ to 1016, which can be represented by an 11-bit signed number. Without the 128 offset, the DC coefficient would range from 0 to 2040, and the other 63 of DCT coefficients would be signed (due to the cosine range). The subtraction of 128 from the pixel block has no effect upon the 63 AC coefficients (an equivalent method could be to perform subtraction of $-1024$ of DC coefficient after the DCT).

## 13.3 Quantization Tables

The quantization table used has a great influence upon the quality of JPEG compression as it influences the degree of compression. These tables are often developed empirically, to give the greatest number of bits to the DCT values which are most noticeable and have the most visible impact.

The quantization table is applied to the output of the DCT, which is an $8 \times 8$ array. The upper left coefficient is the DC coefficient, and the remaining are the 63 AC coefficients of increasing horizontal and vertical frequencies as one moves rightward and downward. As the human eye is more sensitive to lower frequencies, less quantization and more bits are used for the upper and leftmost DCT coefficients.

Example baseline tables are provided in the JPEG standard, as shown in the Table 13.1 below:

| Luminance Quantization Table | | | | | | | |
|---|---|---|---|---|---|---|---|
| 16 | 11 | 10 | 16 | 24 | 40 | 51 | 61 |
| 12 | 12 | 14 | 19 | 26 | 58 | 60 | 55 |
| 14 | 13 | 16 | 24 | 40 | 57 | 69 | 56 |
| 14 | 17 | 22 | 29 | 51 | 87 | 80 | 62 |
| 18 | 22 | 37 | 56 | 68 | 109 | 103 | 77 |
| 24 | 35 | 55 | 64 | 81 | 104 | 113 | 92 |
| 49 | 64 | 78 | 87 | 103 | 121 | 120 | 101 |
| 72 | 92 | 95 | 98 | 112 | 100 | 103 | 99 |
| Chrominance Quantization Table | | | | | | | |
| 17 | 18 | 24 | 47 | 99 | 99 | 99 | 99 |
| 18 | 21 | 26 | 66 | 99 | 99 | 99 | 99 |
| 24 | 26 | 56 | 99 | 99 | 99 | 99 | 99 |
| 47 | 66 | 99 | 99 | 99 | 99 | 99 | 99 |
| 99 | 99 | 99 | 99 | 99 | 99 | 99 | 99 |
| 99 | 99 | 99 | 99 | 99 | 99 | 99 | 99 |
| 99 | 99 | 99 | 99 | 99 | 99 | 99 | 99 |
| 99 | 99 | 99 | 99 | 99 | 99 | 99 | 99 |

Many other quantization tables claiming greater optimization for the human visual range have been developed for various JPEG versions.

The quantized output array B is formed as follows:

$B_{j,k}$ = rounded $(A_{j,k} / Q_{j,k})$ for j = {0.7}, k = {0.7}

where $A_{j,k}$ is the DCT output array value, $Q_{j,k}$ is the quantization table value.

Examples would be:

Luminance DCT output value of $(A_{0,0}) = 426.27$

$B_{0,0}$ = round $(A_{0,0} / Q_{0,0})$ = round $(426.27 / 16) = 27$

Chrominance DCT output value of $(A_{6,2}) = -40.10$

$B_{6,2}$ = round $(A_{6,2} / Q_{6,2})$ = round $(-40.10 / 99) = 0$

Few values of the output $B_{j,k}$ are possible when the quantization value is high. Using a quantization value of 99, the rounded output values can only be $-1$, 0 or $+1$. In many cases, especially when j or k is three or larger, the $B_{j,k}$ will be rounded to zero, indicating little high-frequency in the image region.

This is lossy compression, so called because data is lost in quantization, and cannot be recovered. The principle is to compress by discarding only data that has little impact on the image quality.

## 13.4 Entropy Coding

The next step is to sequence the quantized array values $B_{j,k}$ as in the order shown in Figure 13.1. The first value $B_{0,0}$ is the quantized DC coefficient. All the subsequent values are AC values.

The entropy encoding scheme is fairly complex. The AC coefficients are coded differently than the DC coefficient. The output of the quantizer often contains many zeros, so special symbols are provided. One is an EOB (end of block) symbol, used when the remaining values from the quantizer are all zero. This allows the encoding to be terminated when the rest of the quantized values are zero. The coded symbols also allow the zero run-length following a non-zero symbol to be specified. This efficiently takes advantage of the zeros present in the quantizer output. This is known as run length encoding.

The DC coefficients are differentially coded across the image blocks. There is no relationship between the DC and AC coefficients. However, DC coefficients in different blocks are likely to be correlated, as adjacent $8 \times 8$ image blocks are likely to have a similar DC or average luminance and chrominance. So only the delta, or difference, is coded for the next DC coefficient, relative to the previous DC coefficient.

Four Huffman code tables are provided in the baseline JPEG standard:

DC coefficient, luminance.

AC coefficients, luminance.

DC coefficient, chrominance.

AC coefficients, chrominance.

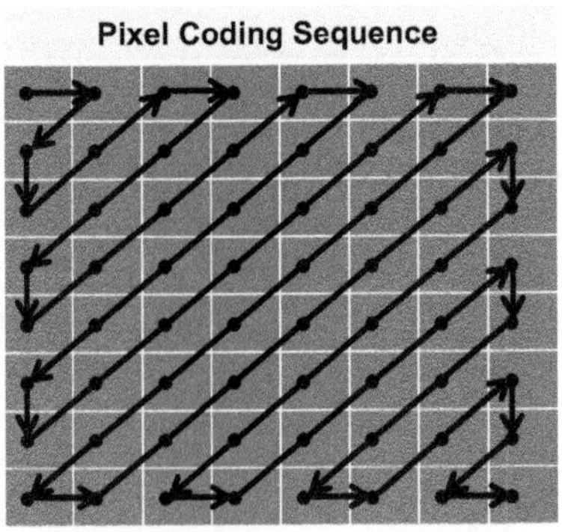

**Pixel Coding Sequence**

**Figure 13.1.** Sequencing of pixel coding.

These tables give encoding for both individual values, and values plus a given number of zeros. Following the properties of Huffman coding, the tables are constructed so that the most statistically common input values are coded using the fewest number of bits. The Huffman symbols are then concatenated into a bit stream that forms the compressed image file. The use of variable length coding makes recovery difficult if any data corruption occurs. Therefore, special symbols or markers are inserted periodically to allow the decoder to resynchronize if there are bit errors in the JPEG file.

The JPEG standard specifies the details of the entropy encoding followed by Huffman coding: it is quite detailed and is not included in this text. For non-baseline JPEG, alternate coding schemes may be used.

For those planning to implement a JPEG encoder or decoder, the following book is recommended: *JPEG Digital Image Compression Standard*, by William Pennebaker and Joan Mitchell.

We have described the various steps in JPEG encoding. The Baseline JPEG process can be summarized by the following encode and decode steps, as shown in Figure 13.2.

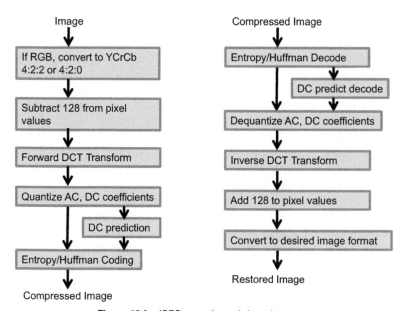

**Figure 13.2.** JPEG encode and decode steps.

## 13.5 JPEG Extensions

The JPEG standard provides for several extensions, some of which are summarized below.

Huffman coding is popular, and has no intellectual property restrictions, but some variants of JPEG use an alternate coding method known as arithmetic coding. Arithmetic coding is more efficient, adapting to changes in the statistical estimates of the input data stream and is subject to patent limitations.

Variable quantization is an enhancement to the quantization procedure of DCT output. This enhancement can be used with the DCTs in JPEG except for the baseline JPEG. The quantization values can be redefined prior to the start of an image scan but must not be changed once they are within a scan.

In this method, the quantization values are scaled at the start of each $8 \times 8$ block — matching the scale factors used to the AC coefficients stored in the compressed data. Quantization values may then be located and changed as needed, which allows for variable quantization based on the characteristics of an image. The variable quantizer continually adjusts during decoding to provide higher quality at the expense of increasing the size of the JPEG file. Conversely, the maximum size of the resulting JPEG file can be set by constant adaptive adjustments made by the variable quantizer.

Another extension is selective refinement, which selects a given region of an image for further enhancement. The resolution of this region of the image is improved using three methods of selective refinement: progressive, hierarchical and component.

Progressive selective refinement is used only in the progressive modes to add more bit resolution of near zero and non-zero DCT coefficients in the region of the image. Hierarchical selective refinement is used in JPEG hierarchical coding mode, and permits for a region of an image to be refined by the next differential image in a defined hierarchical sequence. It allows higher quality or resolution in a given region of the image. Component selective refinement permits a region of a frame to contain fewer colors than are originally defined.

Image tiling is an enhancement that divides a single image into smaller sub-images, which allows for smaller memory buffers, quicker access in both volatile and disk memory and the storing and compression of very large images. There are three types of tiling: simple, pyramidal, and composite.

Simple tiling divides an image into multiple fixed-size tiles. All simple tiles are coded from top to bottom, left to right, and are adjacent. The tiles are all the same size, and encoded using the same procedure.

Pyramidal tiling also partitions the image into multiple tiles, but each tile can have different levels of resolution, resulting in a multi-resolution pyramidal JPEG image. This is known as the JPEG Tiled Image Pyramid (JTIP) model. The JTIP image has successive layers of the same image, but using different resolutions. The top of the pyramid has an image that is one-sixteenth of the defined screen size. It is called the vignette and it can be used for quick displays of image contents. The next image is one-fourth of the screen and is called the imagette — this is often used to display multiple images simultaneously. Next is a lower-resolution, full-screen image and after that are higher-resolution images. The last image is the original image. Each of the pyramidal images can be JPEG encoded, either separately or together in the same data stream. If done separately, then it can allow for faster access of the selected image quality.

Multiple-resolution versions of images can also be stored and displayed using composite tiling, known as a mosaic. Composite tiling differs from pyramidal tiling in three ways: the tiles can overlap, be different sizes, and be encoded using different quantization scaling. Each tile is encoded independently, so they can be easily combined.

Other JPEG extensions are detailed in the JPEG standards.

# 14

# VIDEO COMPRESSION FUNDAMENTALS

## CHAPTER OUTLINE

14.1 Block Size   112
14.2 Motion Estimation   114
14.3 Frame Processing Order   116
14.4 Compressing I-frames   117
14.5 Compressing P-frames   118
14.6 Compressing B-frames   119
14.7 Rate Control and Buffering   119
14.8 Quantization Scale Factor   120

MPEG is often considered synonymous with image compression. MPEG stands for Moving Pictures Experts Group, a committee that publishes international standards on video compression. MPEG is a portfolio of video compression standards, which will be discussed further in the next chapter.

Image compression theory and implementation focuses on taking advantage of the spatial redundancy present in the image. Video is composed of a series of images, usually referred to as frames, and so can be compressed by compressing the individual frames as discussed in the last chapter. However, there are temporal (or across time) redundancies present across video frames: the frame immediately following has a lot in common with the current and previous frames. In most videos there will be a significant amount of repetition in the sequences of frames. This property can be used to reduce the amount of data used to represent and store a video sequence. In order to take advantage of the temporal redundancy, this commonality must be determined across the frames.

This is known as predictive coding, and is effective in reducing the amount of data that must be stored or streamed for a given video sequence. Usually only part of an image changes from frame to frame, which permits prediction from previous frames. Motion compensation is used in the predictive process. If an

Digital Video Processing for Engineers. http://dx.doi.org/10.1016/B978-0-12-415760-6.00014-3

image sequence contains moving objects, then the motion of these objects within the scene can be measured, and the information used to predict the location of the object in frames later in the sequence.

Unfortunately, this is not as simple as just comparing regions in one frame to another. If the background is constant, and objects move in the foreground, there will be significant unchanging areas of the frames. But, if the camera is panning across a scene, there will be areas of subsequent frames that are the same, but will have shifted location from frame to frame. One way to measure this is to sum up all the absolute differences (without regard to sign), pixel by pixel, between two frames. Then the frame can be offset by one or more pixels, and the same comparison run. After many such comparisons, the sum of differences in the results can be compared, where the minimum result corresponds to the best match, and provides the basis for a method to determine the location offset of the match between frames. This is known as the *minimum absolute differences* (MAD) method, or sometimes it's referred to as *sum of absolute differences* (SAD).

An alternative is the *minimum mean square error* (MMSE) method, which measures the sum of the squared pixel differences. This method can be useful because it accentuates large differences, due to the squaring, but the trade-off is the requirement for multiplies as well as subtractions.

# 14.1 Block Size

A practical trade-off has to be made when determining the block size (or macroblock) over which the comparison is run. If the block size is too small there is no benefit in trying to reuse pixel data from a previous frame rather than just use the current frame data. If the block is too large, many more differences will be likely, and it is difficult to get a close match. One obvious block size to run comparisons across would be the $8 \times 8$ block size used for DCT. Experimentation has shown that a $16 \times 16$ pixel area works well, and this is commonly used.

The computational efforts of running these comparisons can be enormous. For each macroblock, 256 pixels must be compared against a $16 \times 16$ area of the previous frame. If we assume the macroblock data could shift by up to, for example, 256 pixels horizontally or 128 pixels vertically (a portion of an HD $1080 \times 1920$) from one frame to another, there will be 256 possible shifts to each side, and 128 possible shifts up or down. This is a total of $512 \times 256 = 131{,}072$ permutations to check, each requiring 256 difference computations. This is over 33 million per macroblock, with $1080 / 16 \times 1920/16 = 8100$ macroblocks per HD image or frame.

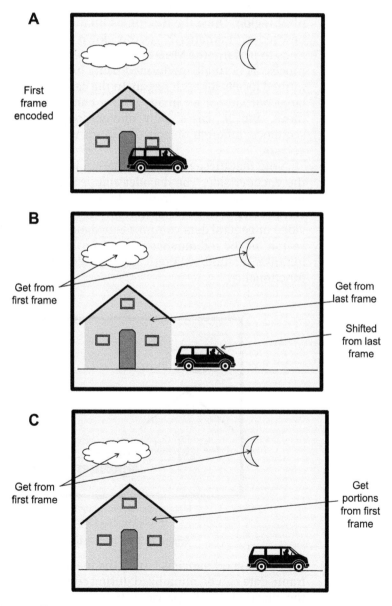

**A**

First frame encoded

**B**

Get from first frame

Get from last frame

Shifted from last frame

**C**

Get from first frame

Get portions from first frame

**Figure 14.1a — 14.1c.** Macroblock matching across nearby frames.

## 14.2 Motion Estimation

Happily, there are techniques to reduce this brute force method. A larger block size can be used initially, over larger portions of the image, to estimate the motion trend between frames. For example, we can enlarge the block size used for motion estimation to include macrocells on all sides, for a $48 \times 48$ pixel region. Once the motion estimation is completed, this can be used to dramatically narrow down the number of permutations to perform the MAD or MMSE over. This motion estimation process is normally performed using luminary (Y) data and performed locally for each region. In the case of the camera panning, large portions of an image with a common motion vector will move, but in cases with moving objects within the video sequence, different objects or regions will have different motion vectors.

Note that these motion estimation issues are present only on the encoder side. The decoder side is much simpler, as it must recreate the image frames based upon data supplied in compressed form. This asymmetry is common in audio, image, video or general data compression algorithms. The encoder must search for the redundancy present in the input data using some iterative method, whereas the decoder does not require this functionality.

Motion estimation vector between
middle and last frame

**Figure 14.2.** Motion estimation.

Once a good match is found, the macroblocks in the following frame data can be minimized during compression by referencing the macroblocks in previous frames. However, even a "good" match will have some error. These errors are referred to as residuals, or residual artifacts. These artifacts can be determined by differences in the four $8 \times 8$ DCTs in the macroblock and then coded and compressed as part of the frame data. Of course, if the residual is too large, coding the residual might require more data compared to just compressing the image data without any reference to the previous frame.

To begin MPEG compression, the source video is first converted to 4:2:0 format, so the chrominance data frame is one fourth the number of pixels — ½ vertical and ½ horizontal resolution. The video must be in progressive mode: each frame is composed of pixels from the same time instant (i.e. not interlaced).

At this point, a bit of terminology and hierarchy used in video compression needs to be introduced:

A **pixel block** refers to an $8 \times 8$ array in a frame.

A **macrocell** is 4 blocks, making a $16 \times 16$ array in a frame.

A **slice** is a sequence of adjacent macrocells in a frame. If data is corrupted, the decoding can typically begin again at the next slice boundary.

A **group of pictures** (GOP) is from one to several frames. The significance of the GOP is that it is self-contained for compression purposes. No frame within one GOP uses data from a frame in another GOP for compression or decompression. Therefore, each GOP must begin with an I-frame (defined below).

A video is made up of a sequence of GOPs.

Most video compression algorithms have three types of frames:

**I-frames** — These are intra-coded frames — they are compressed using only information in the current frame. A GOP always begins with an I-frame, and no previous frame information is required to compress or decompress an I-frame.

**P-frames** — These are predicted frames. P-frames are compressed using image data from an I- or P-frame (that may not be the immediately preceding frame) when compared to the current P-frame. Restoring or decompressing the frame requires compressed data from a previous I- or P-frame, and residual data and motion estimation corresponding to the current P-frame are used. The video compression term for this is inter-coded, meaning the coding uses information across multiple video frames.

**B-frames** — These are bi-directional frames. B-frames are compressed using image data from preceding and successive I- or P-frames. This is compared to the current B data to form the motion estimation and residuals. Restoring or decompressing the B-frame requires compressed data from preceding and successive I- or P-frames, plus residual data and motion estimation corresponding to the current B-frame. B-frames are inter-coded.

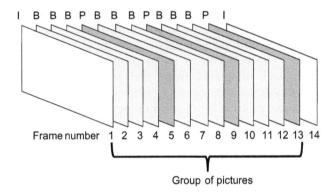

**Figure 14.3.** Group of pictures (GOP) example.

By finishing the GOP with a P-frame, the last few B-frames can be decoded without needing information from the I-frame of the following GOP.

## 14.3 Frame Processing Order

The order of frame encode processing is not sequential. Forward and backward prediction means that each frame has a specific dependency mandating the processing order, and requires buffering of the video frames to allow out-of-sequential order processing. This also introduces multi-frame latency in the encoding and decoding process.

At the start of the GOP, the I-frame is processed first. Then the next P-frame is processed, as it needs the current frame plus information from the previous I- or P-frame. Then the B-frames in between are processed, as, in addition to the current frame, information from both previous and post frames are used. Then the next P-frame, and after that the intervening B-frames, as shown in the processing order given in Figure 14.4.

Note that since this GOP begins with an I-frame and finishes with a P-frame, it is completely self-contained, which is advantageous when there is data corruption during playback or decoding. A given corrupted frame can only impact one GOP. The following GOP, since it begins with an I-frame, is independent of problems in a previous GOP.

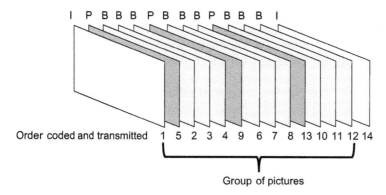

Order coded and transmitted  1  5  2  3  4  9  6  7  8  13 10 11 12 14

Group of pictures

**Figure 14.4.** Video frame sequencing example.

# 14.4 Compressing I-frames

I-frames are compressed in very similar fashion to the JPEG techniques covered in the last chapter. The DCT is used to transform pixel blocks into the frequency domain, after which quantization and entropy coding is performed. I-frames do not use information in any other video frame, therefore an I-frame is compressed and decompressed independently from other frames. Both the luminance and chrominance are compressed, separately. Since the 4:2:0 format is used, the chrominance will have only ¼ of the macrocells compared to the luminance. The quantization table (14.1) used is given below. Once more, notice how larger amounts of quantization are used for higher

## Luminance and Chrominance Quantization Table

| | | | | | | | |
|---|---|---|---|---|---|---|---|
| 8 | 16 | 19 | 22 | 26 | 27 | 29 | 34 |
| 16 | 16 | 22 | 24 | 27 | 29 | 34 | 37 |
| 19 | 22 | 26 | 27 | 29 | 34 | 34 | 38 |
| 22 | 22 | 26 | 27 | 29 | 34 | 37 | 40 |
| 22 | 26 | 27 | 29 | 32 | 35 | 40 | 48 |
| 26 | 27 | 29 | 32 | 35 | 40 | 48 | 58 |
| 26 | 27 | 29 | 34 | 38 | 46 | 56 | 69 |
| 27 | 29 | 35 | 38 | 46 | 56 | 69 | 83 |

horizontal and vertical frequencies, as human vision is less sensitive to high frequency.

Further scaling is provided by a quantization scale factor, which will be discussed further in the rate control description section.

The DC coefficients are coded differentially, using the difference from the previous frame, which takes account of the high degree of the average, or DC, level of adjacent blocks. Entropy encoding similar to JPEG is used. In fact, if all the frames are treated as I-frames, this is pretty much the equivalent of JPEG compressing each frame of image independently.

## 14.5 Compressing P-frames

When processing P-frames, a decision needs to be made for each macroblock, based upon the motion estimation results. If the search for a match with another macroblock does not yield a good match in the previous I- or P-frame, then it must be coded as I-frames are coded — no temporal redundancy can be taken advantage of. However, if a good match is found in the previous I- or P-frame, then the current macroblock can be represented by a motion vector to the matching location in the previous frame, and by computing the residual, quantizing and encoding (inter-coded). The residual uses a uniform quantizer, and this quantizer treats the DC component the same as the rest of the AC coefficients.

## Residual Quantization Table

| | | | | | | | |
|---|---|---|---|---|---|---|---|
| 16 | 16 | 16 | 16 | 16 | 16 | 16 | 16 |
| 16 | 16 | 16 | 16 | 16 | 16 | 16 | 16 |
| 16 | 16 | 16 | 16 | 16 | 16 | 16 | 16 |
| 16 | 16 | 16 | 16 | 16 | 16 | 16 | 16 |
| 16 | 16 | 16 | 16 | 16 | 16 | 16 | 16 |
| 16 | 16 | 16 | 16 | 16 | 16 | 16 | 16 |
| 16 | 16 | 16 | 16 | 16 | 16 | 16 | 16 |
| 16 | 16 | 16 | 16 | 16 | 16 | 16 | 16 |

Some encoders also compare the number of bits used to encode the motion vector and residual to ensure there is a saving when using the predictive representation. Otherwise, the macroblock can be intra-coded.

## 14.6 Compressing B-frames

B-frames try to find a match using both preceding and following I- or P-frames for each macroblock. The encoder searches for a motion vector resulting in a good match over both the previous and following I- or P-frame, and if found uses that frame to inter-code the macroblock. If unsuccessful, then the macroblock must be intra-coded. Another option is to use the motion vector − but with both preceding and following frames simultaneously. When computing the residual, the B macroframe pixels are subtracted to the average of the macroblocks in the two I- or P-frames. This is done for each pixel, to compute the 256 residuals, which are then quantized and encoded.

Many encoders then compare the number of bits used to encode the motion vector and residual using the preceding, following, or average of both frames to see which provides the most bit savings with predictive representation. Otherwise, the macroblock is intra-coded.

## 14.7 Rate Control and Buffering

Video rate control is an important part of image compression. The decoder has a video buffer of fixed size, and the encoding process must ensure that this buffer never under- or overruns the buffer size, as either of these events can cause noticeable discontinuities for the viewer.

The video compression and decompression process requires processing of the frames in a non-sequential order. Normally, this is done in real-time. Consider these three scenarios:

Video storage − video frames arriving at 30 frames per minute must be compressed (encoded) and stored to a file.

Video readback − a compressed video file is decompressed (decoded), producing 30 video frames per minute.

Streaming video − a video source at 30 frames per minute must be compressed (encoded) to transmit over a bandwidth-limited channel (only able to support a maximum number of bits/second throughput). The decompressed (decoded) video is to be displayed at 30 frames per minute at the destination.

All of these scenarios will require buffering, or temporary storage of video files during the encoding and decoding process. As mentioned above, this is due to the out-of-order processing of the video frames. The larger the GOP, containing longer sequences of I-, P- and B-frames, the more buffering is potentially needed.

The number of bits needed to encode a given frame depends upon the video content. A video frame with lots of constant background (the sky, a concrete wall) will take few bits to encode, whereas a complex static scene like a nature film will require many more bits. Fast moving, blurring scenes with a lot of camera panning will be somewhere in-between as the quantizing of the high frequency DCT coefficients will tend to keep the number of bits moderate.

I-frames take the most bits to represent, as there are no temporal redundancy benefits. P-frames take fewer bits, and the least bits are needed for B-frames, as B-frames can leverage motion vectors and predictions from both previous and following frames. P and B-frames will require far fewer bits than I-frames if there is little temporal difference between successive frames, or conversely, may not be able to save any bits through motion estimation and vectors if the successive frames exhibit little temporal correlation.

Despite this, the average rate of the compressed video stream often needs to be held constant. The transmission channel carrying the compressed video signal may have a fixed bit-rate, and keeping this bit-rate low is the reason for compression in the first place. This does not mean the bits for each frame, for example at a 30 Hz rate, will be equal, but that the average bit rate over a reasonable number of frames may need to be constant. This requires a buffer, to absorb higher numbers of bits from some frames, and provide enough bits to transmit for those frames encoded with few bits.

## 14.8 Quantization Scale Factor

Since the video content cannot be predicted in advance, and this content will require a variable amount of bits to encode the video sequence, provision is made to dynamically force a reduction in the number of bits per frame. A scale factor is applied to the quantization process of the AC coefficients of the DCT, which is referred to as Mquant.

The AC coefficients are first multiplied by eight, then divided by the value in the quantization table, and then divided again by Mquant. Mquant can vary from 1 to 31, with a default value of eight. For the default level, Mquant just cancels the initial multiplication of eight.

One extreme is an Mquant of one. In this case, all AC coefficients are multiplied by 8, then divided by the quantization table value. The results will tend to be larger, non-zero numbers, which will preserve more frequency information at the expense of using more bits to encode the slice. This results in higher quality frames.

The other extreme is an Mquant of 31. In this case, all AC coefficients are multiplied by eight, then divided by the quantization table value, then divided again by 31. The results will tend to be small, mostly zero numbers, which will remove most spatial frequency information and reduce bits to encode the slice resulting in lower quality frames. Mquant provides a means to trade quality versus compression rate, or number of bits to represent the video frame. Mquant normally updates at the slice boundary, and is sent to the decoder as part of the header information.

This process is complicated by the fact that the human visual system is sensitive to video quality, especially for scenes with little temporal or motion activity. Preserving a reasonably consistent quality level needs to be considered as the Mquant scale factor is varied.

Here the state of the decoder input buffer is shown. The buffer fills at a constant rate (positive slope) but empties discontinuously as various frame size I, P, and B data is read for each frame decoding process. The amount of data for each of these frames will vary according to video content and the quantization scale factors that the encode process has chosen.

To ensure the encoder process never causes the decode buffer to over- or underflow, it is modeled using a video buffer verifier

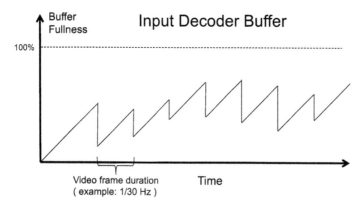

**Figure 14.5.** Video frame buffer behavior.

(VBV) in the encoder. The input buffer in the decoder is conservatively sized, due to consumer cost sensitivity. The encoder will use the VBV model to mirror the actual video decoder state, and the state of the VBV can be used to drive the rate control algorithm, which in turn dynamically adjusts the quantization scale factor used in the encode process and is sent to the decoder. This closes the feedback loop in the encoder to ensure the decoder buffer does not over- or underflow.

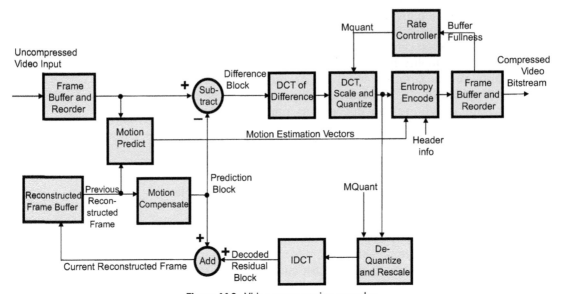

**Figure 14.6.** Video compression encoder.

The video encoder is shown as a simplified block diagram. The following steps take place in the video encoder:

1. Input video frames are buffered and then ordered. Each video frame is processed macroblock by macroblock.

2. For P- or B-frames, the video frame is compared to an encoded reference frame (another I- or P-frame). The motion estimation function searches for matches between macroblocks of the current and previously encoded frames. The spatial offset between the macroblock position in the two frames is the motion vector associated with macroblock.

3. The motion vector points to be best-matched macroblock in the previous frame – called a motion-compensated prediction macroblock. It is subtracted from the current frame macroblock to form the residual or difference.

4. The difference is transformed using the DCT and quantized. The quantization also uses a scaling factor to regulate the average number of bits in compressed video frames.

5. The quantizer output, motion vector and header information is entropy (variable length) coded, resulting in the compressed video bitstream.

6. In a feedback path, the quantized macroblocks are rescaled and transformed using the IDCT to generate the same difference or residual as the decoder. It has the same artifacts as the decoder due to the quantization processing, which is irreversible.

7. The quantized difference is added to the motion-compensated prediction macroblock (see step two above). This is used to form the reconstructed frame, which can be used as the reference frame for the encoding of the next frame. It is worth remembering that the decoder will only have access to reconstructed video frames, not the actual original video frames, to use reference frames.

The video decoder is shown as a simplified block diagram. The following steps take place in the video decoder:

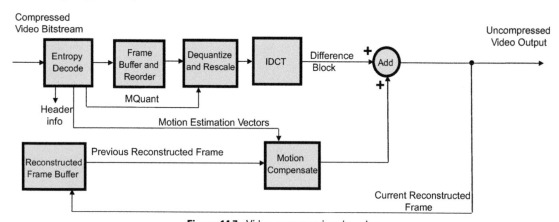

**Figure 14.7.** Video compression decoder.

1. Input compressed stream entropy decoded. This extracts header information, coefficients, and motion vectors.

2. The data is ordered into video frames of different types (I, P,B), buffered and re-ordered.

3. For each frame, at the macroblock level, coefficients are rescaled and transformed using IDCT, to produce the difference, or residual block.

4. The decoded motion vector is used to extract the macroblock data from a previous decoded video frame. This

becomes the motion-compensated prediction block for the current macroblock.

5. The difference block and the motion-compensated prediction block are summed together to produce the reconstructed macroblock. Macroblock by macroblock, the video frame is reconstructed.

6. The reconstructed video frame is buffered, and can be used to form prediction blocks for subsequent frames in the GOP. The buffered frames are output from the decoder in sequential order for viewing.

The decoder is, in essence, a subset of the encoder functions, and fortunately requires much fewer computations. This supports the often asymmetric nature of video compression. A video source can be compressed once, using higher-performance broadcast equipment, and then distributed to many users, who can decode using much less costly consumer-type equipment.

# 15

# FROM MPEG TO H.264 VIDEO COMPRESSION

## CHAPTER OUTLINE

**15.1 MPEG-2   126**

15.1.1 MPEG-2 Levels and Profiles   126

   *15.1.1.1 MPEG-2 Support for 4:2:2   128*

   *15.1.1.2 MPEG-2 Interlaced Video Support   129*

   *15.1.1.3 MPEG-2 Frame and Field DCT   129*

   *15.1.1.4 MPEG-2 frame and field motion prediction and estimation   129*

   *15.1.1.5 MPEG-2 Quantization   129*

   *15.1.1.6 MPEG-2 support for 3:2 pulldown   130*

   *15.1.1.7 MPEG-2 Aspect Ratio Support   130*

**15.2 H.264 Video Compression Standard   131**

15.2.1 H.264 Levels and Profiles   131

   *15.2.1.1 H.264 Profiles   131*

15.2.2 H.264 Support for 4 × 4 Integer DCT   135

15.2.3 H.264 Logarithmic Quantization   135

15.2.4 H.264 Frequency Dependent Quantization   135

15.2.5 H.264 Pixel Interpolation   136

15.2.6 H.264 Variable Block Size Motion Estimation   136

15.2.7 H.264 Deblocking Filter   137

15.2.8 H.264 Temporal Prediction   137

15.2.9 H.264 Motion Adaptive Frame and Field Coding   138

15.2.10 H.264 Spatial Prediction   138

15.2.11 H.264 Entropy Coding   138

15.2.12 H.264 Redundant Slices   139

15.2.13 H.264 Arbitrary Slice Ordering   139

15.2.14 H.264 Flexible Macroblock Ordering   139

15.2.15 H.264 Additional Features   139

**15.3 Digital Cinema Applications   140**

MPEG was introduced in the previous chapter, and we know that MPEG is a group containing several standards, of which we described the MPEG-1 version. MPEG-2 followed, and is by far the most popular of the MPEG standards: for example, DVDs contain video compressed using MPEG-2.

Digital Video Processing for Engineers. http://dx.doi.org/10.1016/B978-0-12-415760-6.00015-5

>As there was no release of MPEG-3, the next standard was MPEG-4. In parallel, the Video Coding Experts Group (VCEG) — a committee operating under the ITU organization — published video compression standards. They developed the H.261 and H.263 standards, which were optimized for video conferencing, in which low bit-rates and low latencies are particularly important.

These two groups eventually merged into the Joint Video Team (JVT). VCEG had started work on an Advanced Video Codec (AVC), which was similar to the methods used in MPEG-4 version 10. The merged efforts have become known as H.264 part 10/AVC, or H.264/AVC, or just H.264. Blu-ray disc video is compressed using H.264, as is much of the video content streamed from YouTube.

This chapter will begin by highlighting the enhancements from MPEG-1 to MPEG-2, which help explain the widespread adoption of MPEG-2. MPEG-4, in its early versions, wasn't as widely adopted as it did not offer enough improvement over MPEG-2 to warrant changing. There were also some disputes over licensing terms. H.264 offers a more substantial level of improvement and flexibility over MPEG-2. Due to the prevalence of high definition video, it has been widely adopted in studios, consumer video players and by internet video-streaming sites. The latest standard, H.265, has still not been finalized as of this writing, and is not covered here.

## 15.1 MPEG-2

The techniques of the previous chapter form the basis of all the subsequent video compression standards. I-, P- and B-frames are still used. In MPEG-2, it is typical though not required for every 15$^{th}$ frame to be compressed as an I-frame.

Despite the common technical base with MPEG-1, there are important differences and improvements in MPEG-2, and many of these will be reviewed here.

### 15.1.1 MPEG-2 Levels and Profiles

A level is used to constrain resources in the decoder, which often must meet strict cost limitations. By establishing different levels, the decoder manufacturer can choose how much video processing resources (and cost) to provide to meet a particular capability. The level setting puts a limit on parameters such as the maximum frame size, the frame rate, size of decoder video-frame buffer and how many previous frames may be used for prediction. For example, a fairly low level might be used on a mobile device, while a higher level would be used on home entertainment equipment. The mobile device has a much smaller resolution screen, will have greater constraints on video bit rates, and is very cost sensitive. The home entertainment system will most likely be a 1080p system. The common MPEG-2 levels are low, main, and high. There is also a level for a particular European high-definition format with the frame aspect ratio of 1440 × 1152 pixels.

# Table 15.1
## MPEG-2 Levels

| Abbr. | Name | Frame rates (Hz) | Max horizontal resolution | Max vertical resolution | Max luminance samples per second (approximately height x width x framerate) | Max bit rate in Main profile (Mbit/s) |
|---|---|---|---|---|---|---|
| LL | Low Level | 23.976, 24, 25, 29.97, 30 | 352 | 288 | 3,041,280 | 4 |
| ML | Main Level | 23.976, 24, 25, 29.97, 30 | 720 | 576 | 10,368,000, except in High profile, where constraint is 14,475,600 for 4:2:0 and 11,059,200 for 4:2:2 | 15 |
| H-14 | High 1440 | 23.976, 24, 25, 29.97, 30, 50, 59.94, 60 | 1440 | 1152 | 47,001,600, except that in High profile with 4:2:0, constraint is 62,668,800 | 60 |
| HL | High Level | 23.976, 24, 25, 29.97, 30, 50, 59.94, 60 | 1920 | 1152 | 62,668,800, except that in High profile with 4:2:0, constraint is 83,558,400 | 80 |

Profiles are groups of algorithmic options that may be used in the decoding process and allow different levels of complexity in decoder implementations. A decoder used in a broadcast studio environment will probably be more sophisticated than that used in home equipment.

Due to the new features in the permissible profiles, MPEG-2 was able to accommodate a wide range of compression quality requirements, compared to the rather simple Mquant adjustment

# Table 15.2
## MPEG-2 Example Profiles

| Abbr. | Name | Picture Coding Types | Chroma Format | Aspect Ratios | Scalable modes | Intra DC Precision |
|---|---|---|---|---|---|---|
| SP | Simple profile | I, P | 4:2:0 | square pixels, 4:3, or 16:9 | none | 8, 9, 10 |
| MP | Main profile | I, P, B | 4:2:0 | square pixels, 4:3, or 16:9 | none | 8, 9, 10 |
| SNR | SNR Scalable profile | I, P, B | 4:2:0 | square pixels, 4:3, or 16:9 | SNR (signal-to-noise ratio) scalable | 8, 9, 10 |
| Spatial | Spatially Scalable profile | I, P, B | 4:2:0 | square pixels, 4:3, or 16:9 | SNR- or spatial-scalable | 8, 9, 10 |
| HP | High profile | I, P, B | 4:2:2 or 4:2:0 | square pixels, 4:3, or 16:9 | SNR- or spatial-scalable | 8, 9, 10, 11 |
| 422 | 4:2:2 profile | I, P, B | 4:2:2 or 4:2:0 | square pixels, 4:3, or 16:9 | none | 8, 9, 10, 11 |
| MVP | Multi-view profile | I, P, B | 4:2:0 | square pixels, 4:3, or 16:9 | Temporal | 8, 9, 10 |

used in MPEG-1. The trade-off is increased encoder and decoder complexity and processing rates. This, plus the use of levels, allows for standardized video-compression implementation of various video resolutions and features.

Combinations of profiles and levels can be used. There are many possible permutations defined in the MPEG-2 standard — the most common profile used is the main profile, which defines the set of algorithms or decoding tools most frequently used.

# Table 15.3
# Common MPEG-2 Level and Profile Permutations

| Profile @ Level | Resolution (px) | Framerate max. (Hz) | Sampling | Bitrate (Mbit/s) | Example Application |
|---|---|---|---|---|---|
| SP@LL | 176 × 144 | 15 | 4:2:0 | 0.096 | Wireless handsets |
| SP@ML | 352 × 288 | 15 | 4:2:0 | 0.384 | PDAs |
| | 320 × 240 | 24 | | | |
| MP@LL | 352 × 288 | 30 | 4:2:0 | 4 | Set-top boxes (STB) |
| MP@ML | 720 × 480 | 30 | 4:2:0 | 15 (DVD: 9.8) | DVD, SD-DVB |
| | 720 × 576 | 25 | | | |
| MP@H-14 | 1440 × 1080 | 30 | 4:2:0 | 60 (HDV: 25) | HDV |
| | 1280 × 720 | 30 | | | |
| MP@HL | 1920 × 1080 | 30 | 4:2:0 | 80 | ATSC 1080i, 720p60, HD-DVB (HDTV). (Bitrate for terrestrial transmission is limited to 19.39Mbit/s) |
| | 1280 × 720 | 60 | | | |
| 422P@LL | | | 4:2:2 | | |
| 422P@ML | 720 × 480 | 30 | 4:2:2 | 50 | Sony IMX using I-frame only, Broadcast "contribution" video (I&P only) |
| | 720 × 576 | 25 | | | |
| 422P@H-14 | 1440 × 1080 | 30 | 4:2:2 | 80 | |
| | 1280 × 720 | 60 | | | |
| 422P@HL | 1920 × 1080 | 30 | 4:2:2 | 300 | Sony MPEG HD422 (50 Mbit/s), Canon XF Codec (50 Mbit/s), Convergent Design Nanoflash recorder (up to 160 Mbit/s) |
| | 1280 × 720 | 60 | | | |

Some of the features below are defined as part of profiles that extend beyond the main profile.

### 15.1.1.1 MPEG-2 Support for 4:2:2

In MPEG-1, compression was performed on video in the 4:2:0 format. In MPEG-2, a 4:2:2 profile now allows the source video to be in 4:2:2, which is commonly used in broadcast studio processing. This allows for compression in the higher-quality chroma-resampling format.

### 15.1.1.2 MPEG-2 Interlaced Video Support

As discussed in earlier chapters, interlaced video is a legacy of television broadcast systems, used to raise the frame-update rate to avoid flicker while not increasing the actual data rate. Interlaced video is challenging for compression, because the alternating rows are from different time instants and can have different motion vectors, so that the optimum predictive match may be from different frames. MPEG-2 adds techniques to cope with interlaced video, some of which are listed below.

### 15.1.1.3 MPEG-2 Frame and Field DCT

MPEG-1 uses a frame-type DCT; that is the DCT is performed on $8 \times 8$ arrays of pixels in the same frame. This would be sub-optimal when using interlaced video, as differences between the interlaced pixel rows in the frame could introduce substantial high-frequency content into the DCT coefficients, leading to poor compression.

MPEG-2 has an option to use field DCT. When using field DCT, each of the four $8 \times 8$ blocks in the $16 \times 16$ macrocell array has the DCT performed independently, resulting in four DCT coefficient arrays of $8 \times 8$ each. For interlaced video, the $16 \times 16$ macrocell is divided into two 8 wide by 16 tall pixel arrays. An $8 \times 8$ DCT transform is performed on the odd rows (1, 3, 5, 7) of each $8 \times 8$ pixel array. This results in the top array of $8 \times 8$ DCT coefficients. A second $8 \times 8$ DCT transform is then performed on the even rows (2, 4, 6, 8) of each $8 \times 8$ pixel array, and is used to compute the bottom array of $8 \times 8$ DCT coefficients.

In this way, the interlaced sets of rows, which correspond to the two different instants in time of the video frame, are transformed and quantized separately.

### 15.1.1.4 MPEG-2 frame and field motion prediction and estimation

Progressive video uses the $16 \times 16$ macroblock to perform prediction and motion estimation. When using the field DCT it also makes sense to perform the motion estimation using each $16 \times 8$-sized pixel array corresponding to each field of the interlaced video macroblock.

### 15.1.1.5 MPEG-2 Quantization

Additional bits can be used for DCT coefficients, beyond the eight bits used in MPEG-1. Up to 10 bits can be used.

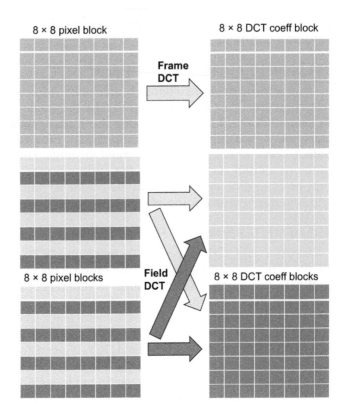

**Figure 15.1.** Frame and field DCT.

### 15.1.1.6 MPEG-2 support for 3:2 pulldown

MPEG-2 allows for compression of video in the 3:2 pulldown form. Film or movie video is commonly recorded at 24 fps, progressively. For use in television, it's decoded into 60 fps. This is done using 3-2 pulldown. Each film frame is replicated either two or three times, resulting in 60 fps.

24 fps film video sequence: A, B, C, D....

60 fps 3:2 pulldown sequence: A, A, B, B, B, C, C, D, D, D....

This is an optional output format which the decoder can provide, to ensure compatibility with televisions or monitors.

### 15.1.1.7 MPEG-2 Aspect Ratio Support

Video recorded and compressed with 16:9 aspect ratio can be decoded to display the portion of frame corresponding to the 4:3 aspect ratio used in older televisions and monitors.

## 15.2 H.264 Video Compression Standard

This trend of more profiles and levels continues in H.264. The complexity and processing requirements increase dramatically as H.264 can compress approximately twice as much video as MPEG-2, at similar quality levels. Alternatively, H.264 can deliver substantially better quality at the same bit rate as MPEG-2.

Many of the key H.264 enhancements are listed below, followed by a more detailed discussion of some of these enhancements. For those interested in deeper discussion of the implementation details and options on H.264, there are a number of H.264 texts written at a level for video engineers, and of course, there is the H.264 standard itself.

### 15.2.1 H.264 Levels and Profiles

The levels and profiles have been expanded to enable new options. Table 15.4 summarizes these. The levels provide for maximum permissible constraints for various parameters, which drive the processing power, memory and other requirements of the encoder and especially decoder.

The profiles include baseline constrained, baseline, extended baseline, main and various high-quality broadcast-level profiles. Some of the features defined in the profile description are explained in the following sections.

#### 15.2.1.1 H.264 Profiles

Constrained Baseline Profile (CBP)

Mainly found in low-cost applications, this profile is most typically used in videoconferencing and mobile applications. It has the minimum feature set. It was released after the Baseline profile, and is intended to further eliminate features not required in specific applications, such as redundant slices, arbitrary slice-ordering and flexible macroblock ordering.

Baseline Profile (BP)

This profile is used primarily in low-cost applications where there may be data loss during transmission, such as mobile devices or video conferencing. It is similar to CBP, but has features to help compensate for data loss or corrupted frames. Baseline profile is also intended for low latency applications. It includes I- and P-frames, but excludes B-frames, which require more latency to support the frame buffering required by backwards prediction. Baseline profiles use the $4 \times 4$ DCT transform

# Table 15.4
# H.264 Levels

| Level | Max macroblocks | | Max video bit rate (video coding layer – VCL) | | | | Examples for high resolution @ frame rate (max stored frames) |
|---|---|---|---|---|---|---|---|
| | per second | per frame | BP, XP, MP (kbit/s) | HiP (kbit/s) | Hi10P (kbit/s) | Hi422P, Hi444PP (kbit/s) | |
| 1 | 1,485 | 99 | 64 | 80 | 192 | 256 | 128×96@30.9 (8)<br>176×144@15.0 (4) |
| 1b | 1,485 | 99 | 128 | 160 | 384 | 512 | 128×96@30.9 (8)<br>176×144@15.0 (4) |
| 1.1 | 3,000 | 396 | 192 | 240 | 576 | 768 | 176×144@30.3 (9)<br>320×240@10.0 (3)<br>352×288@7.5 (2) |
| 1.2 | 6,000 | 396 | 384 | 480 | 1,152 | 1,536 | 320×240@20.0 (7)<br>352×288@15.2 (6) |
| 1.3 | 11,880 | 396 | 768 | 960 | 2,304 | 3,072 | 320×240@36.0 (7)<br>352×288@30.0 (6) |
| 2 | 11,880 | 396 | 2,000 | 2,500 | 6,000 | 8,000 | 320×240@36.0 (7)<br>352×288@30.0 (6) |
| 2.1 | 19,800 | 792 | 4,000 | 5,000 | 12,000 | 16,000 | 352×480@30.0 (7)<br>352×576@25.0 (6) |
| 2.2 | 20,250 | 1,620 | 4,000 | 5,000 | 12,000 | 16,000 | 352×480@30.7(10)<br>352×576@25.6 (7)<br>720×480@15.0 (6)<br>720×576@12.5 (5) |
| 3 | 40,500 | 1,620 | 10,000 | 12,500 | 30,000 | 40,000 | 352×480@61.4 (12)<br>352×576@51.1 (10)<br>720×480@30.0 (6)<br>720×576@25.0 (5) |
| 3.1 | 108,000 | 3,600 | 14,000 | 17,500 | 42,000 | 56,000 | 720×480@80.0 (13)<br>720×576@66.7 (11)<br>1280×720@30.0 (5) |
| 3.2 | 216,000 | 5,120 | 20,000 | 25,000 | 60,000 | 80,000 | 1,280×720@60.0 (5)<br>1,280×1,024@42.2 (4) |
| 4 | 245,760 | 8,192 | 20,000 | 25,000 | 60,000 | 80,000 | 1,280×720@68.3 (9)<br>1,920×1,080@30.1 (4)<br>2,048×1,024@30.0 (4) |
| 4.1 | 245,760 | 8,192 | 50,000 | 62,500 | 150,000 | 200,000 | 1,280×720@68.3 (9)<br>1,920×1,080@30.1 (4)<br>2,048×1,024@30.0 (4) |
| 4.2 | 522,240 | 8,704 | 50,000 | 62,500 | 150,000 | 200,000 | 1,920×1,080@64.0 (4)<br>2,048×1,080@60.0 (4) |
| 5 | 589,824 | 22,080 | 135,000 | 168,750 | 405,000 | 540,000 | 1,920×1,080@72.3 (13)<br>2,048×1,024@72.0 (13)<br>2,048×1,080@67.8 (12)<br>2,560×1,920@30.7 (5)<br>3,680×1,536@26.7 (5) |
| 5.1 | 983,040 | 36,864 | 240,000 | 300,000 | 720,000 | 960,000 | 1,920×1,080@120.5 (16)<br>2,560×1,920@51.2 (9)<br>4,096×2,048@30.0 (5)<br>4,096×2,304@26.7 (5) |
| 5.2 | 2,073,600 | 36,864 | 240,000 | 300,000 | 720,000 | 960,000 | 1,920×1,080@172.0 (16)<br>2,560×1,920@108.0 (9)<br>4,096×2,048@63.3 (5)<br>4,096×2,304@56.3 (5) |

to reduce computational load on the decoder. It supports CAVLC entropy coding.

### Main Profile (MP)

This is optimized for broadcast applications, with support for interlaced video. It also has features for weighted predictions, meaning the predicted frame made up of proportions of the values from several frames. This is used to efficiently compress fades and dissolves in the video. All three frame types – I, P and B are supported – as well as features for interlaced video. It supports the CABAC entropy coding.

### Extended Profile (XP)

This profile is optimized for internet streaming applications, where both high compression rates and the ability to cope with data loss is important. It's able to tolerate data losses during transmission or switching. Some additional features are included to make it easier to interface to the streaming network layer.

### High Profile (HiP)

Intended for high-definition video applications, high profiles are used on video disks such as Blu-ray. The high profiles are a superset of the main profile, and support extended bit widths, higher chroma-sampling options, frequency-dependent quantization and use of the higher computational $8 \times 8$ DCT transform.

### Progressive High Profile (PHiP)

This profile is used in high-definition video applications, but does not have any support for interlaced video features.

### High 10 Profile (Hi10P)

Professional broadcast studios use high 10 profiles to process high-definition video. This uses 10 bits per sample, rather than the eight bits often used in consumer products.

### High 4:2:2 Profile (Hi422P)

Similar to the Hi10P, Hi422P adds support for 4:2:2 chroma-sampled video, and is mainly used in professional broadcast studios.

### High 4:4:4 Predictive Profile (Hi444PP)

Again for professional broadcast applications, using up to 4:4:4 chroma sampling Hi444PP allows up to 14 bits per sample. The color planes can optionally be coded separately.

### High 10 Intra Profile

The High 10 Profile with intra-only use (no B- or P-frames permitted). This is necessary when the video is edited in professional studios.

### High 4:2:2 Intra Profile

The High 4:2:2 Profile with intra-only use.

### High 4:4:4 Intra Profile

The High 4:4:4 Profile with intra-only use.

### CAVLC 4:4:4 Intra Profile

The High 4:4:4 Profile with intra-only use and only CAVLC coding.

The H.264 standard is too complex for an introductory book to cover, however we will here cover some of the major enhancements or features of H.264 (see Table 15.5).

## Table 15.5
## H.264 Profile Features

| Feature | CBP | BP | XP | MP | HiP | Hi10P | Hi422P | Hi444PP |
|---|---|---|---|---|---|---|---|---|
| Chroma formats | 4:2:0 | 4:2:0 | 4:2:0 | 4:2:0 | 4:2:0 | 4:2:0 | 4:2:0/4:2:2 | 4:2:0/4:2:2/4:4:4 |
| Sample depths (bits) | 8 | 8 | 8 | 8 | 8 | 8 to 10 | 8 to 10 | 8 to 14 |
| Flexible macroblock ordering (FMO) | No | Yes | Yes | No | No | No | No | No |
| Arbitrary slice ordering (ASO) | No | Yes | Yes | No | No | No | No | No |
| Redundant slices (RS) | No | Yes | Yes | No | No | No | No | No |
| Data Partitioning | No | No | Yes | No | No | No | No | No |
| SI and SP slices | No | No | Yes | No | No | No | No | No |
| B slices | No | No | Yes | Yes | Yes | Yes | Yes | Yes |
| Interlaced coding (PicAFF, MBAFF) | No | No | Yes | Yes | Yes | Yes | Yes | Yes |
| CABAC entropy coding | No | No | No | Yes | Yes | Yes | Yes | Yes |
| 8×8 vs. 4×4 transform adaptivity | No | No | No | No | Yes | Yes | Yes | Yes |
| Quantization scaling matrices | No | No | No | No | Yes | Yes | Yes | Yes |
| Separate $C_b$ and $C_r$ QP control | No | No | No | No | Yes | Yes | Yes | Yes |
| Monochrome (4:0:0) | No | No | No | No | Yes | Yes | Yes | Yes |
| Separate color plane coding | No | No | No | No | No | No | No | Yes |
| Predictive lossless coding | No | No | No | No | No | No | No | Yes |

## 15.2.2 H.264 Support for 4 × 4 Integer DCT

In a 4 × 4 DCT, each transform coefficient has a contribution from 16 pixels, compared to the 8 × 8 DCT, where each coefficient has a contribution from 64 pixels. Therefore the 4 × 4 DCT has less computations per pixel.

One of the key elements in H.264 is the ability to compress video using a 4 × 4 DCT on 4 × 4 pixel blocks. In addition, many of the DCT coefficients and operations are optimized for digital arithmetic. Since the DCT output coefficients are to be quantized afterwards, the precision of the DCT calculation can be reduced. For example, coefficients which are close to 0.5 are approximated to 0.5. This does not require a multiplier, but can be implemented with a right-shift operation. Furthermore, some divides can be eliminated by pre-multiplying the input data by factors of two, which can be done with left shifts. These steps can allow all coefficients to be simple integers. The net result is a major simplification in the DCT, which is where a significant portion of the processing takes place.

The same simplification process is applied to the 8 × 8 DCT used in H.264. However, not all of the multiplies and divides can be eliminated in the 8 × 8 DCT, due to its greater dynamic range.

The simplifications are of particular help in parallel hardware implementations, such as FPGAs and ASSPs, where the multiplier elimination can save significant circuit area.

## 15.2.3 H.264 Logarithmic Quantization

The quantization in the earlier MPEG standards was linear. In H.264 the Mquant parameter is replaced by the QP (quantization parameter), which has a value from 0 to 51. However, the effect of QP is not linear. Each increase of six in the QP value results in a doubling of the quantization step size. The logarithmic scaling provides finer step sizes for small QP, and coarser step sizes for large QP. The result is a greater dynamic range for scaling, with the same number of bits to represent the QP parameter in the quantization process.

## 15.2.4 H.264 Frequency Dependent Quantization

Human vision is less sensitive to artifacts when there are complex scenes, which tend to have high spatial frequencies. Previously, the quantization applied to the DCT transform coefficients has been uniform, meaning all coefficients use the same quantization rules. Frequency-dependent quantization allows the quantization step to change depending on the position of the transform coefficient. Bigger steps, and more quantization, are applied to the coefficients representing higher spatial frequencies.

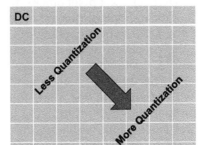

**Figure 15.2.** Frequency-dependent DCT transform coefficient quantization.

## 15.2.5 H.264 Pixel Interpolation

Motion vectors specify the location of a matching frame. If the match is not close, coding of the residual will require more bits than just coding the frame without any prediction. To assist in matching, the motion vectors in MPEG-2 allow matching to a pixel or half-pixel resolution. To perform half-pixel matching, the linear interpolation is used − that is the pixel in-between is calculated as the average of the four adjacent pixels side-to-side, top and bottom. H.264 improves upon this in two ways. First, it provides a better interpolation, executing the half-pixel interpolation using the three adjacent pixels on each side. Second, it then allows for a further quarter-pixel interpolation, using a linear interpolation between the nearest half-pixels. Quarter pixel resolution for matching enables a better predictive match to be made, with smaller residuals, resulting in higher compression.

## 15.2.6 H.264 Variable Block Size Motion Estimation

In the MPEG-2 standard, motion estimation is always made through matching of $16 \times 16$ macrocells. In H.264, block sizes can vary from $16 \times 16$ to $4 \times 4$, enabling more accurate segmentation of moving regions of the frame. For luma prediction, block sizes include $16 \times 16$, $16 \times 8$, $8 \times 16$, $8 \times 8$, $8 \times 4$, $4 \times 8$, and $4 \times 4$. These can be used in any combination within a single macroblock. For chroma-prediction blocks, the sizes may be smaller depending upon the chroma subsampling in use (4:4:4 or 4:2:2 or 4:2:0).

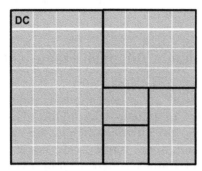

**Figure 15.3.** Macroblock decomposition example for motion estimation.

## 15.2.7 H.264 Deblocking Filter

One of the main artifacts of MPEG-2 was blocking, or being able to detect boundaries between the macrocell boundaries. Due to quantization in the DCT process, many of the macroblocks might have the same DCT coefficient values. Sometimes, nearby blocks have sufficient difference in the pixels to cross the threshold in the quantization process. When the differing DCT coefficients become visually apparent, the edges of the macroblocks start to appear. This is especially a problem at high compression ratios where the quantization is very coarse.

H.264 requires a deblocking filter. The deblocking filter will smooth or filter pixel values at the macroblock boundaries for a better visual appearance. Due to the large numbers of macroblocks and edges involved, this can impose a high computational load. Various techniques are used to mitigate the number of computations.

## 15.2.8 H.264 Temporal Prediction

The P- and B-frames, as in MPEG-2, are predictive frames. P-frames utilize an earlier frame, while B-frames use both an earlier and later frame. In H.264, multiple frames (up to a maximum of 16) can be used for prediction. The frames selected are identified in a reference picture index. This can allow better prediction, at the expense of much higher computations for the encoder. Of course, the use of multiple reference frames for prediction requires the ability to weigh, or scale the contribution from the different frames. This provides a more accurate prediction, especially in cases with repetitive motion, or in the situation where an object is present in one frame and then absent in the next, suddenly uncovering the static background scenery.

However, most B-frames will still use only two predictive frames, as in MPEG-2.

## 15.2.9 H.264 Motion Adaptive Frame and Field Coding

When using MPEG-2, a field mode is defined using $16 \times 8$ half macroblocks. In H.264, macroblock-adaptive frame-field (MBAFF) coding, $16 \times 16$ macroblocks are used in the field mode. Picture-adaptive frame-field coding (PAFF) uses a combination of frames coded as progressive or interlaced frames, depending upon local motion, similar to motion-adaptive deinterlacing. This can allow individual frames, or even macroblocks, to be encoded as interlaced or progressive, depending on whether there are significant differences between the top and bottom fields.

## 15.2.10 H.264 Spatial Prediction

Within a given frame, H.264 allows for a spatial intra-prediction from the edges of neighboring blocks. The macroblock can be predicted from other nearby macroblocks either above or to the left. Spatial prediction of $4 \times 4$ blocks is also permitted in the same manner. Note that MPEG-2 only allowed for DC coefficient prediction between nearby frames.

## 15.2.11 H.264 Entropy Coding

Two alternatives were developed to Huffman or variable length coding. One is content-based adaptive variable-length coding (CAVLC). This more efficient because the coding is in fixed tables, but adaptively changes with the data composition, or context. The compression process of predicting, applying DCT transforms, and quantization produces many zeros. Run-length encoding is used to compactly represent these zero strings. After zero, the most likely results to be coded are $+/-1$.

There are several tables used to perform the variable length coding, selected based on the data, or content. The $8 \times 8$ blocks are broken into four $4 \times 4$ blocks, each of which can use a different table, depending upon the content.

Another method used in some profiles is context-based adaptive binary coding (CABAC). CABAC compresses data even more efficiently than CAVLC but requires more processing power to decode.

## 15.2.12 H.264 Redundant Slices

In lossy channels, redundant slices can be added, for all or part of a frame. If there are no transmission errors, the decoder will ignore the redundant slices. If a slice is corrupted, the decoder can use the redundant slice instead. However, the addition of redundant slices increases the bit rate and reduces the compression rate.

## 15.2.13 H.264 Arbitrary Slice Ordering

Slices are normally sequentially ordered from left to right and from top to bottom. They are also transmitted in this order. Arbitrary slice ordering (ASO) allows the slices to be transmitted in any order. By transmitting out of order, and in some interleaved fashion, the effect of errors can sometimes be reduced. In a packetized system, the loss or corruption of a packet might affect several slices. Normally, these would be adjacent slices, but with ASO they can be spread out throughout the frame, minimizing the visual impact of the packet error.

## 15.2.14 H.264 Flexible Macroblock Ordering

Extending the same idea as ASO, flexible macroblock ordering can be used to further offset the effect of errors. Macroblocks are normally coded sequentially in raster order within a slice. However, it is possible to assign macroblocks to other slices, thereby interleaving the macroblocks across the various slices of a frame. There is a macroblock allocation map that describes different options for distribution of macroblocks to the different slice groups.

In the event of a corrupted slice, the macroblocks of that slice are then in different regions of the frame, according to the macroblock allocation map. Since the neighbors of the damaged macroblock from the corrupted slice are intact, the damaged macroblock can be concealed by interpolating from the intact neighboring macroblocks.

## 15.2.15 H.264 Additional Features

There are many more features in various H.264 profiles, which cannot be covered in any detail here. For example, numerous features are provided to support streaming operation; for recovery when frame data is corrupted or missing; to better interface with network stacks; there is a Network Abstraction Layer (NAL) syntax defined to allow H.264 to be used in many network environments; and packetization is supported, with minimal header overhead. Two sets of high-level parameters are defined as the Sequence Parameter Set (SPS) and the Picture Parameter Set (PPS); these more advanced features are not discussed further in this text.

## 15.3 Digital Cinema Applications

Digital cinema is the use of digital compression technology in the motion picture industry, including areas such as: filming or recording; storage and distribution; and playing or projecting in movie theaters using digital representation and compression. Digital cinema can be much higher resolution than that for home entertainment systems, and therefore demands correspondingly higher storage requirements when distributing and archiving a full-length movie.

Movies can be made with 1080p (1920 × 1080) pixel resolution. The next frame size is known as 2K for the vertical resolution, using 2048 × 1080. 4K increases the resolution to 4096 × 2160 pixels per frame. The latest professional cameras can record at 8K or 8196 × 4320 pixels per frame, in 3-D form.

High-end movie cameras also sample much faster — typically 120 fps. The production video can then be fairly easily converted to 24, 48, or 60 fps. For high speed or motion applications, the video could be left in 120 frames per second.

Theater projector and playback equipment costs are a key factor in the adoption of digital cinema. Traditional film projectors may cost ~$50,000, and can last for 30 years or more. A digital-cinema system costs in the order of $150,000 or more, and since it's mostly computer-based electronic equipment, is more difficult and expensive to repair. It is also likely to have a shorter lifetime, and perhaps be obsolete within five years. The benefit, of course, is the screen resolution, quality and technical advancements such as 3-D enabled by digital cinema technology.

To encourage adoption, some of the theater costs may be subsidized by the film producers and distributors, as digital cinema allows for the electronic distribution of films and more extensive anti-piracy measures, which save the industry money. Still, there is little doubt that the digital technology adoption trend is contributing to higher prices for the movie theater customer.

# 16

# VIDEO NOISE AND COMPRESSION ARTIFACTS

## CHAPTER OUTLINE

16.1 Salt-and-pepper Noise    141
16.2 Mosquito Noise    143
16.3 Block Artifacts    144

Video or image data can contain noise: this can be introduced by the sensor, during transmission or during storage. Imperfections on old film can cause noise when it's converted to digital video form — this tends to be a random type of noise. Noise reduction processing or filtering is a process that can reduce or eliminate this effect.

## 16.1 Salt-and-pepper Noise

Salt-and-pepper noise occurs when individual pixels or small areas of the image have high contrast, or are very different in color or intensity from their surrounding pixels or areas. This pixel or area is visually noticeable and means that the image contains dark and white dots — which is why it's referred to as salt-and-pepper noise. Sources include particles of dust inside the camera and defective CCD elements.

Filtering can be employed to counter this type of noise. A linear filter with a low-pass characteristic can be employed to smooth sharp transitions in both vertical and horizontal directions. This type of filtering is very effective, but it will remove, or at least reduce the amount of, all high-frequency spatial content. Thus sharp edges will also be smoothed out with the resultant blurring, or loss of sharpness of the image.

Alternatively, another type of filter, known as the median filter, is often used. A conventional digital filter is a linear combination, or weighting of nearby pixels, for the new adjusted value for a given pixel. The right combination of weighting, or coefficients, results in a smoothing, or low pass, effect.

Digital Video Processing for Engineers. http://dx.doi.org/10.1016/B978-0-12-415760-6.00016-7

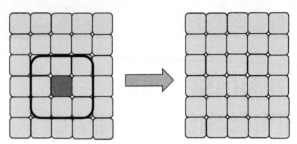

**Figure 16.1.** Median filter of random noise.

The median filter is a non-linear filter, and, unlike a conventional digital filer, does not require any multipliers. The median filter operates by taking a group of pixels in a certain area, for example a 3 × 3 pixel area. The new pixel value in the center of that area is the median, or middle value, of the nine pixel values. There is no weighting or linear combination. The entire image is processed by shifting the 3 × 3 pixel area in vertical and horizontal increments. If a single pixel is very different from all surrounding pixels, it will never be the median value, and will be filtered out by the median filter. This is shown in the black and white pixel image in Figure 16.1.

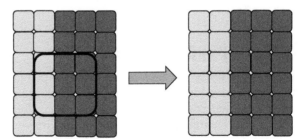

**Figure 16.2.** Median filter with sharp edge.

However, consider a sharp vertical edge: this will be smoothed or softened by a low pass digital filter. But with a median filter, the same sharp transition can be preserved, as shown in Figure 16.2.

Alternate pixel areas can be used for median filtering, as shown in Figure 16.3. The operations are the same — the output pixel value at the center of the pixel area is the median, set equal to the value of the pixel value in the middle after all the pixels have been sorted by value. As the median filter is shifted across the image, it always operates on the original pixel data. The output of the median filter forms a new image, built up pixel by pixel using the median filter. The median filter is but one type of filter that can be used. There are other types of more complex filtering that can be applied.

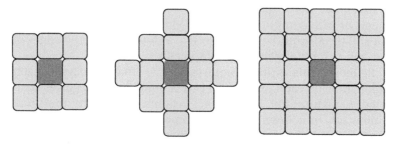

**Figure 16.3.** Examples of median filter area.

Removal of salt-and-pepper noise also has a beneficial effect upon subsequent video compression. This type of noise creates high frequency, which needs to be encoded using high-frequency DCT coefficients, requiring portions of the compressed bit-stream to be wasted in representing this noise.

## 16.2 Mosquito Noise

Video compression and decompression can result in other types of video distortion or artifacts. One common distortion happens near sharp or crisp edges of objects in MPEG and other video frames that use the discrete cosine transform (DCT). It occurs during decompression when the decoding engine has to approximate the discarded data by inverting the transform model. Known as mosquito noise, it manifests as random aliasing in these areas and appears as ringing around sharp edges. It is caused by the removal of high-frequency coefficients during compression quantization, as shown in Figure 16.4. As TVs and monitors get larger, mosquito noise and other artifacts become more prominent.

**From EETimes article "Video compression artifacts and MPEG noise reduction article" by Algolith, 2006**

**Figure 16.4.** Mosquito noise example.

Mosquito noise tends to occur around text or letters in an image, or with computer-generated video objects with very sharp transitions. Mosquito noise looks like a cloud around the edges of text and computer generated graphics.

Reducing this artifact requires spatial image analysis. Each region, or even each pixel, must be classified into categories such as edge, texture, flat and artifact regions. Across the temporal dimension, or sequential video frames, motion prediction can also be used to help distinguish moving edges due to objects in motion. A combination of this analysis is used to identify artifact regions and apply filtering techniques to reduce in those regions. This tends to be an adaptive process.

Sometimes the only choice may be to accept the noise, or to filter and have blurring occur. When motion is very rapid, filtering may be the better choice, even if blurring is introduced, as the blurring is often not as obvious in fast motion scenes. And on some occasions, the motion is too rapid for the temporal rate (frame per second), violating the sampling theorem, so that some form of degradation due to the aliasing is inevitable.

## 16.3 Block Artifacts

Block artifacts are a discernable visual blocking effect on the video which can occur at the $8 \times 8$ boundaries used in the DCT. This is due to several effects, one of which is differing DC coefficients. In some cases, this effect can become visually objectionable, as shown on the left side of Figure 16.5. Each block can also have varying amounts of quantization, depending upon how many AC coefficients have significant amplitude. When high quantization is used, the DCT averages each $8 \times 8$ block approximately, making it roughly appear like a group of pixels, each of size $8 \times 8$. These effects can become even more pronounced in scenes with high motion, such as sports broadcasts.

A common solution is to filter at the block edges, to smooth the transition between blocks. This filter must by necessity be of short order, and apply to only one or two pixels at the $8 \times 8$ boundary. This filter is adaptively applied and is part of the H.264 decoder in some profiles. One of the factors driving it is the quantization level: at high quantization levels, blocking artifacts are more predominant, and deblocking filtering is needed. Another factor is video content: a smooth region will have a stronger blocking effect compared to an area with a high-activity video level. The smoother region will require more aggressive deblocking filtering.

"Deblocking Filter for Low Bit Rate MPEG-4 Video" by Shen-Chuan Tai, Yen-Yu Chen, and Shin-Feng Sheu

**Figure 16.5.** Deblocking filter example.

The H.264 deblocking filter is designed to avoid the requirement for multipliers, instead choosing coefficients that can be implemented using shift and adds, reducing the hardware requirements.

Figure 16.5 shows the image prior to deblocking, and then after application of the deblocking filter. All of these processes can be applied to individual color planes, or to luma and chroma planes.

# VIDEO MODULATION AND TRANSPORT

## CHAPTER OUTLINE

17.1 Complex Modulation and Demodulation   147
17.2 Modulated Signal Bandwidth   150
17.3 Pulse Shaping Filter   152
17.4 Raised Cosine Filter   155
17.5 Signal Upconversion   163
17.6 Digital Upconversion   164

Video compression is often performed to reduce the data rate, or bandwidth, necessary for transport. Video is sent over satellite, fiber, coax or copper twisted-pair wires. In practice, multiple video signals are often sent over the same medium simultaneously.

## 17.1 Complex Modulation and Demodulation

Modulation is the process of taking information bits and mapping them to symbols.

This sequence of symbols is filtered to produce a baseband waveform with the desired spectral properties. The baseband waveform or signal is then upconverted to a carrier frequency, which can be transmitted over the air, through coaxial cable, through fiber or some other medium. At some point in this process a DAC converts the signal to analog, which is amplified and transmitted.

One of the most common and simplest modulation methods is known as QPSK (quadrature phase shift keying). With QPSK, every two input bits will map to one of four symbols, as shown below in Figure 17.1. The complex plane is used in order to represent two dimensions.

The bitstream of zeros and ones input to the modulator is converted into a stream of symbols. Each symbol is represented

Digital Video Processing for Engineers. http://dx.doi.org/10.1016/B978-0-12-415760-6.00017-9

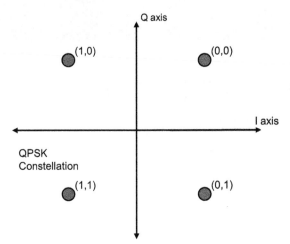

**Figure 17.1.** QPSK constellation.

as the coordinates of a location in the I-Q plane. In QPSK, there are four possible symbols, arranged as shown. Since there are four symbols, the input data are arranged as groups of two bits, which are mapped to the appropriate symbol. The arrangement of symbols in the I-Q plane is also called the constellation.

## Table 17.1 QPSK symbol mapping
### Constellation size and bit rate

| Input Bit Pair | I Value | Q Value | Symbol Value (Location On Complex Plane) |
|---|---|---|---|
| 0, 0 | 1 | 1 | $I + jQ => 1 + j$ |
| 0, 1 | 1 | −1 | $I + jQ => 1 - j$ |
| 1, 0 | −1 | 1 | $I + jQ => -1 + j$ |
| 1, 1 | −1 | −1 | $I + jQ => -1 - j$ |

Another common modulation scheme is known as 16-QAM (quadrature amplitude modulation), which has 16 symbols, arranged as shown in Figure 17.2. Again, don't worry about the name of the modulation. Since we have 16 possible symbols, each symbol will map to four bits. To put it another way, in QPSK each symbol carries two bits of information, while in 16-QAM, each symbol carries four bits of information.

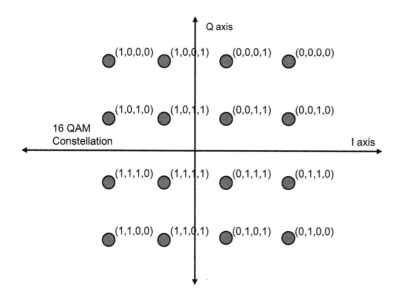

**Figure 17.2.** 16-QAM constellation.

As shown, 16-QAM is the more efficient modulation method. For simplicity, assume symbols are transmitted at a rate of 1 MHz, or 1 MSPS (normally, this is much higher). Then our system, if using the 16-QAM modulation, will be able to send 4 Mbits/second. If instead QPSK is used in this system, it will be able to send only 2 Mbits/second. We could also use 64-QAM which is an even more efficient constellation. Since there are 64 possible symbols, arranged as eight rows of eight symbols each, each symbol carries six bits of information, and supports a data rate of 6 Mbits/second. A few sample constellation types are shown in Table 17.2.

# Table 17.2
## Constellation size and bit rate

| Modulation Type | Possible number of Symbols | Bits per Symbol | Transmitted Bit Rate |
|---|---|---|---|
| QPSK | $4 = 2^2$ | 2 | 2 * symbol rate |
| 8 PSK | $8 = 2^3$ | 3 | 3 * symbol rate |
| 16-QAM | $16 = 2^4$ | 4 | 4 * symbol rate |
| 64-QAM | $64 = 2^6$ | 6 | 6 * symbol rate |
| 256-QAM | $256 = 2^8$ | 8 | 8 * symbol rate |

The frequency bandwidth is primarily determined by the symbol rate. A QPSK signal at 1 MSPS will require about the same bandwidth as a 16-QAM signal at 1 MSPS. Notice the 16-QAM modulator is able to send twice the data within this bandwidth, compared to the QPSK modulator. There is a trade-off however: as the number of symbols increases, it becomes more and more difficult for the receiver to detect which symbol was sent. If the receiver needs to choose from 16 possible symbols which could have been transmitted, rather than choose from from four possibilities, it is more likely to make errors.

The level of errors will depend upon the noise and interference present in the signal, the strength of the signal and how many possible symbols the receiver must select from. In non-line-of-sight systems (cellular phones), there are often high levels of reflected and weak signals due to buildings or other objects blocking the transmission path. In this situation, it is often preferable to use a simple constellation, such as QPSK. Even with a weak signal, the receiver can usually make the correct choice of four possible symbols. Line-of-sight systems, such as satellite systems, have directional receive and transmit antennas facing each other. Because of this, the interfering noise level is usually very low, and complex constellations such as 64-QAM or 256-QAM can be used.

Assuming the receiver is able to make the correct choice from among 64 symbols, three times more bits can be encoded into each symbol, resulting in a $3 \times$ higher data rate. Some adaptive communication systems allow the transmitter to dynamically switch between constellation types depending on the quality of the connection between the transmitter and receiver.

## 17.2 Modulated Signal Bandwidth

We will examine an example of the QPSK constellation, with a transmission rate of 1 MSPS. The baseband signal is two dimensional, so must be represented with two orthogonal components, which are by convention denoted I and Q.

Consider a sequence of five QPSK symbols, at time t = 1, 2, 3, 4 and 5 respectively. The sequence in the two-dimensional complex-constellation plane will appear as a signal trajectory moving from one constellation point to another over time.

The individual I and Q signals are plotted against time in Figure 17.4. This is a two-dimensional signal, with each component plotted separately. The I and Q baseband signals are

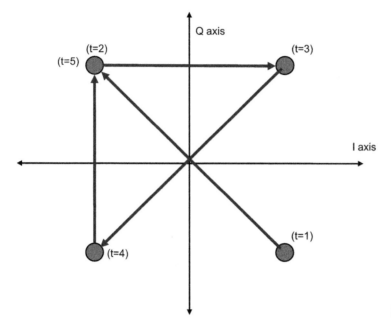

**Figure 17.3.** Constellation trajectory.

generated as two separate signals, and later combined together with a carrier frequency to form a single passband signal.

Note the sharp transitions of the I and Q signals at each symbol. This requires the signal to have high-frequency content. A signal that is of low frequency can change only slowly and smoothly.

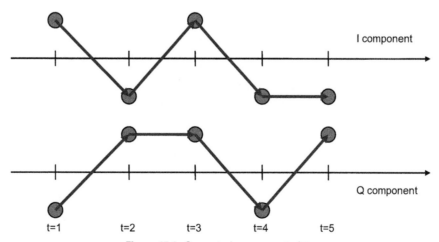

**Figure 17.4.** Separated component plots.

The high-frequency content of these plotted I and Q signals can cause problems, because in most systems it is important to minimize the frequency content, or bandwidth, of the signal. We have already discussed frequency response, where a low-pass filter removes fast transitions or changes in a signal (or eliminates the high-frequency components of that signal). If the frequency response of the signal is reduced, this is the same as reducing its bandwidth. The smaller the bandwidth of the signal, the more signal channels and therefore capacity can be packed into a given amount of frequency spectrum. Thus the channel bandwidth is often carefully controlled.

A simple example is FM radio. Each station, or channel, is located 200 kHz from its neighbor. That means that each station has 200 kHz spectrum, or frequency response, it can occupy. The station on 101.5 is transmitting with a center frequency of 101.5 MHz. The channels on either side transmit with center frequencies of 101.3 and 101.7 MHz. Therefore it is important to restrict the bandwidth of each FM station to within $+/-100$ kHz, to ensures it does not overlap or interfere with neighboring stations. The bandwidth is restricted using a low-pass filter. Video channels are usually several MHz bandwidth, but the same concept applies.

The signal's frequency response, or spectrum, can be shifted up or down the frequency axis at will, using a complex mixer, or multiplier. This is called up- or downconversion.

## 17.3 Pulse Shaping Filter

To accomplish frequency limiting of the modulated signal, the I and Q signals are low-pass filtered. This filter is often called a pulse shaping filter, and it determines the bandwidth of the modulated signal. The filter time domain response is also important. Assuming an ideal low-pass filter is used where symbols are generated at a rate (R) of 1 MSPS. The period T is the symbol duration, and equal to 1 in this example. The relationship between the rate R and symbol period T is:

$R = 1 / T$ and $T = 1 / R$

Alternating with positive and negative I and Q values at each sample interval (this is the worst case in terms of high frequency content), the rate of change will be 500 kHz. So we would start with a low-pass filter with passband of 500 kHz.

This filter will have the sin(x) / x or sinc impulse response. The impulse response is centered in Figure 17-6, along with preceding and following symbol responses. It has zero crossings at intervals

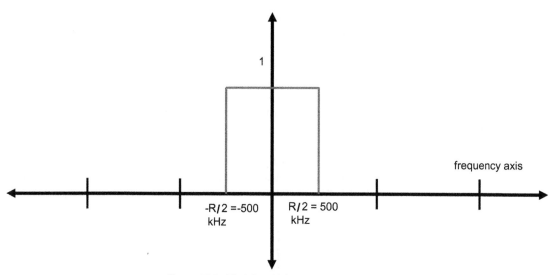

**Figure 17.5.** Modulated signal frequency width.

of T seconds, and decays slowly. A very long filter is needed to approximate the sinc response. The impulse response of the symbols immediately preceding and following the center symbol are in Figure 17.6 shown below. The signal transmitted will be the sum of all the symbols' impulse response (we are just showing three symbols here). If the I or Q sample has a negative value for a particular symbol, then the impulse response for that symbol will be inverted from what is shown below.

The receiver is sampling the signal at T intervals in time to detect the symbol values. It must sample at the T intervals shown on the time axis above (leave aside for now the question of how

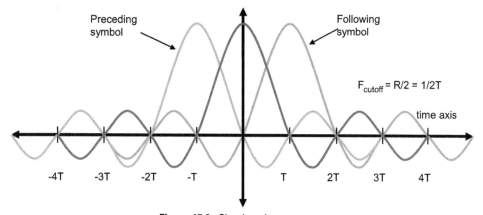

**Figure 17.6.** Sinc impulse response.

the receiver knows exactly where to sample). At time $t = -T$, the receiver will be sampling the first symbol. Notice how the two later symbols have zero crossings at $t = -T$, and so have no contribution at this instant. At $t = 0$, the receiver will be sampling the value of the second symbol. Again, the other symbols, such as first and third adjacent symbols, have zero crossings at $t = 0$, and have no contribution. If the bandwidth of the filter is reduced to less than 500 kHz ($R / 2$) then in the frequency domain these pulses would widen (remember that the narrower the frequency spectrum, the longer the time response, and vice versa). Figure 17.7 shows the result if the $F_{cutoff}$ of the pulse shaping filter is narrowed to 250 kHz, or $R / 4$.

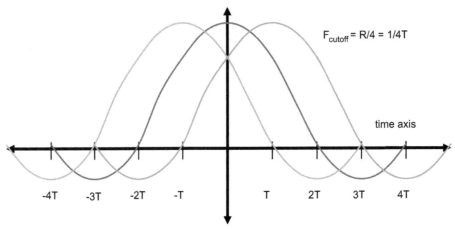

$F_{cutoff} = R/4 = 1/4T$

time axis

-4T   -3T   -2T   -T        T   2T   3T   4T

**Figure 17.7.** Narrow filter impulse response.

In this case, notice how at time $t = 0$, the receiver will be sampling contributions from all three pulses. At each sampling point of t equal to $\ldots -3T, -2T, -T, 0, T, 2T \ldots$ the signal is going to have contributions from many nearby symbols, preventing detection of any specific symbol. This phenomenon is known as inter-symbol interference (ISI), and shows that transmitting symbols at a rate R requires at least R Hz (or $1 / T$ Hz) in the passband frequency spectrum. At baseband, the equivalent two-dimensional (complex) spectrum is from $-R / 2$ to $+R / 2$ Hz) to avoid creating ISI. Therefore to transmit a 1 MSPS signal over the air, at least 1 MHz of RF frequency spectrum will be required. The baseband filters will need a cutoff frequency of at least 500 kHz.

Notice that the frequency spectrum or bandwidth required depends on the symbol rate, ***not*** the bit rate. We can have a much higher bit rate, depending on the constellation type used. For example, each 256-QAM symbol carries eight bits of information, while a QPSK symbol only carries two bits of information. But if they both have a common symbol rate, both constellations require the same bandwidth.

Two issues remain: the sinc impulse response decays very slowly, and so will take a long filter (many multipliers) to implement; and although the response of the other symbols does go to zero at the sampling time when $t = N \cdot T$, where N is any integer, it can be seen visually that if the receiver samples just a little bit to either side, the adjacent symbols will contribute. This makes the receiver symbol-detection performance very sensitive to the sampling timing.

The ideal would be for an impulse response that still goes to zero at intervals of T, but decay faster and have lower amplitude lobes, or tails. This way, when sampling a bit to one side of the ideal sampling point, the lower amplitude tails will make the unwanted contribution of the neighboring symbols smaller. By making the impulse response decay faster, we can reduce the number of taps and therefore multipliers required to implement the pulse shaping filter.

## 17.4 Raised Cosine Filter

There is a type of filter commonly used to meet these requirements. It is called the "raised cosine filter", and it has an adjustable bandwidth, controlled by the "roll off" factor. The trade-off will be that the bandwidth of the signal will become a bit wider, and therefore more frequency spectrum will be required to transmit the signal.

# Table 17.3
## Raised cosine filter frequency rolloff table

| Roll off Factor | Label | Comments — excess bandwidth refers to the percentage of additional bandwidth required compared to ideal low-pass filter |
|---|---|---|
| 0.10 | A | Requires long impulse response (high multiplier resources), has small frequency excess bandwidth of 10% |
| 0.25 | B | A commonly used roll off factor, excess bandwidth of 25% |
| 0.50 | C | A commonly used roll off factor, excess bandwidth of 50% |
| 1.00 | D | Excess bandwidth of 100%, never used in practice |

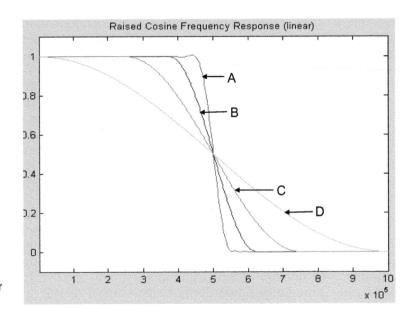

**Figure 17.8.** Raised cosine filter frequency response.

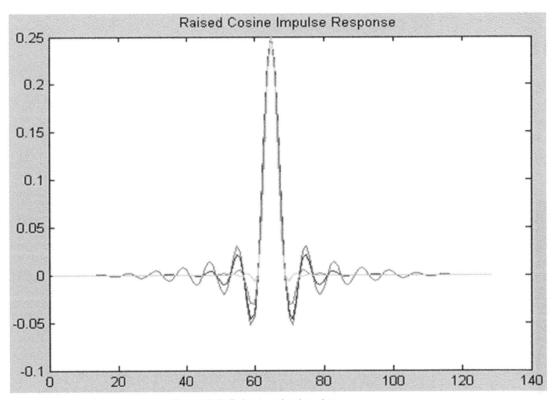

**Figure 17.9.** Raised cosine impulse response.

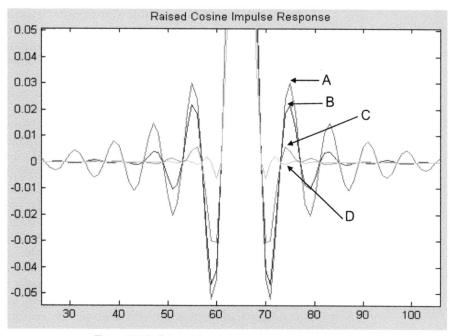

**Figure 17.10.** Zoomed in raised cosine impulse response.

Table 17.3 summarizes the raised cosine response shown in Figure 17.8 for different roll off factors. These labels are also used in Figures 17.9 and 17.10.

In Figure 17.8 the frequency response of the raised cosine filter is shown. To better see the passband shape, it is plotted linearly, rather than logarithmically (dB). It has a cutoff frequency of 500 kHz, the same as our ideal low-pass filter. A raised cosine filter response is wider than the ideal low-pass filter, due to the transition band. This excess frequency bandwidth is controlled by a parameter called the "roll off" factor. The frequency response is plotted for several different roll off factors. As the roll off factor gets closer to zero, the transition becomes steeper, and the filter approaches the ideal low-pass filter.

The impulse response of the raised cosine filter is portrayed in Figure 17.9, which shows the filter impulse response, and Figure 17.10 zooms in to better show the lobes of the filter impulse. Again, it is plotted for several different roll off factors. It is similar to the sinc impulse response in that it has zero crossings at time intervals of T (as this is shown in sample domain, rather than time domain, it is not readily apparent from the diagram).

As the excess bandwidth is reduced to approach the ideal low-pass filter frequency response, the lobes in the impulse response become higher, approaching the sinc impulse response. The signal with the smaller amplitude lobes has a larger excess bandwidth, or wider spectrum.

The pulse shaping filter should have a zero response at intervals of T in time so that a given symbol's pulse response will not have a contribution to the signal at the sampling times of the neighboring symbols. It should also minimize the height of the lobes of the impulse (time) response and have it decay quickly, to reduce the sensitivity to ISI if the receiver doesn't sample precisely at the correct time for each symbol. As the roll off factor increases, this is exactly what happens in the figures (D). The impulse response goes to zero very quickly, and the lobes of the filter impulse response are very small. However, they have a frequency spectrum that is excessively wide. A better compromise would be a roll off factor somewhere between 0.25 and 0.5 (B and C). Here the impulse response decays relatively quickly with small lobes, requiring a pulse shaping filter with a small number of taps, while still keeping the required bandwidth reasonable.

The roll off factor controls the compromise between:
- spectral bandwidth requirement.
- length or number of taps of pulse shaping filter.
- receiver sensitivity to ISI.

Another significant aspect of the pulse shaping filter is that it is always an interpolating filter. In the figures, this is shown as a four times interpolation filter. Looking carefully at the impulse response in Figure 17.10, the zero crossing occurs every four samples. This corresponds to $t = N \cdot T$ in the time domain, due to the $4 \times$ interpolation.

The pulse shaping filter must an interpolating filter, as the I and Q baseband signals must meet the Nyquist criterion. In this example, the symbol rate is 1 MSPS. Using a high roll off factor, the baseband spectrum of the I and Q signals can be as high as 1 MHz. So a minimum sampling rate of 2 MHz, or twice the symbol rate, is required. Therefore, the pulse shaping filter will need to interpolate by at least a factor of two, and is often interpolated significantly higher than this.

Using pulse shaped and interpolated I and Q baseband digital signals, digital-to-analog converters create the analog I and Q baseband signals. These signals can be used to drive an analog mixer which can create a passband signal. A passband signal is a baseband signal upconverted or mixed with a carrier frequency. This process can also be done digitally, with a DAC used at much higher frequency to output the signal at the carrier frequency.

For example, using a 0.25 roll off filter for a 1 MSPS modulator, the baseband I and Q signals will have a bandwidth of 625 kHz. Using a carrier frequency of 1 GHz, the transmit signal will require about 1.25 MHz of spectrum centered at 1 GHz.

Until now the transmission path has been our focus. The receive path is quite similar. The signal is down converted, or mixed down to baseband. The demodulation process starts with baseband I and Q signals. The receiver is more complex, as it must deal with several additional issues. There may be nearby signals which can interfere with the demodulation process, which must be filtered out, usually with a combination of analog and digital filters. The final stage of digital filtering is often the same pulse shaping filter used in the transmitter − called a matched filter. The idea is that the same filter that was used to create the signal is also used to filter the spectrum prior to sampling, maximizing the amount of signal energy used in the detection (or sampling process). Because we are using the same filter in the transmitter and the receiver, the raised cosine filter is usually modified to a square-root raised-cosine filter. The frequency response of the raised cosine filter is modified to be the square root of the amplitude across the passband. This also modifies the impulse response as well. This is shown in the figures below, for the same roll off factors.

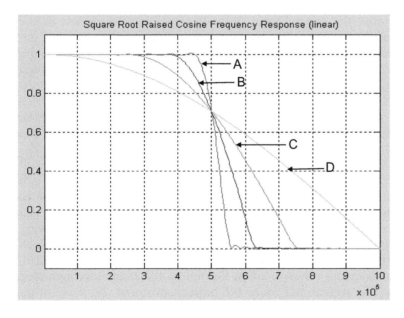

**Figure 17.11.** Square root raised cosine frequency response.

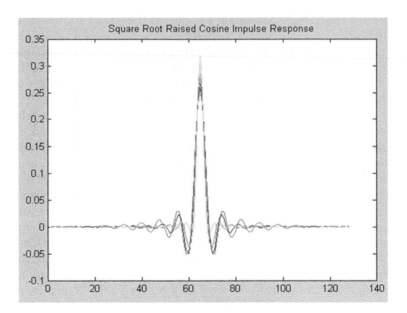

**Figure 17.12.** Square root raised cosine impulse response.

Since the signal passes through both filters, the net frequency response is the raised cosine filter. After passing through the receive pulse shaping (also called matched) filter, the signal is sampled. Using the sampled I and Q value, the receiver will choose the constellation point in the I-Q plane closest to the sampled value. The bits corresponding to that symbol are recovered, and if all went well, the receiver has chosen the same symbol point selected by the transmitter. This is very much simplified, but is the essence of a digital communications process.

It can be seen why it will be easier to have errors when transmitting 64-QAM as compared to QPSK. The receiver has 64 closely spaced symbols to select from in the case of 64-QAM, whereas in QPSK there are only four widely spaced symbols to select from. This makes 64-QAM systems much more susceptible to ISI, noise or interference. An option is to just transmit the 64-QAM signal with higher power, to spread the symbols further apart. This is an effective, but very expensive, way to mitigate the noise and interference which prevents correct detection of the symbol at the receiver. Also, the transmit power is often limited by the regulatory agencies, or the transmitter may be battery powered or have other constraints.

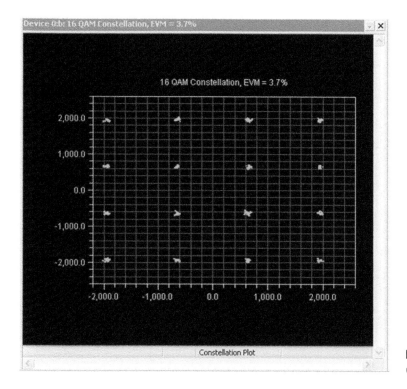

**Figure 17.13.** 16-QAM recovered constellation.

The receiver also has a number of other problems to contend with. The sampling is assumed to be at the correct instant in time when one symbol has a non-zero value in the signal. The receiver must determine this correct sampling time, usually by a combination of trial and error during the initial part of the reception, and sometimes by having the transmitter send a predetermined (or training) sequence known by both transmitter and receiver. This process is known as acquisition, where the receiver tries to fine-tune the sampling time, the symbol rate, the exact frequency and phase of the carrier and other parameters which may be needed to demodulate the received signal with a minimum of errors. And once all this information is determined, it must still be tracked, to account for differences in transmit and receive clocks, or other impairments.

Figures 17.13 and 17.14 are plots from both a 16-QAM and 64-QAM constellation after being sampled by an actual digital receiver – a WiMax wireless system operating in the presence of noise. Each receiver signal has the same average energy. The receiver does manage to do a sufficiently good job at detection so that each of the constellation points is clearly visible. As the

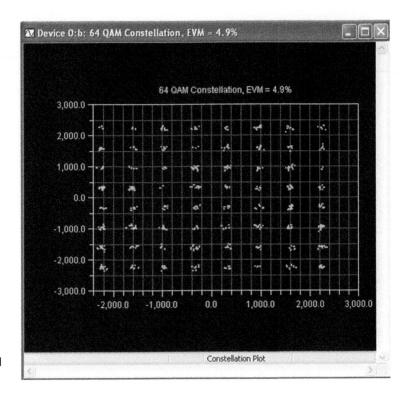

**Figure 17.14.** 64-QAM recovered constellation.

receiver noise level increases, the constellation samples would quickly start to drift together on the 64-QAM constellation, and it would not be possible to accurately determine which constellation point a given symbol should map to. The 16-QAM system is more robust in the presence of additive noise and other impairments than the 64-QAM.

The modulation and demodulation (modem) ideas presented in this chapter are used in most digital communication systems, including satellite, cable and fiber systems used to transmit video signals. In QPSK all four symbols have the same amplitude. The phase in the complex plane is what distinguishes the different symbols, each of which is located in a different quadrant. For QAM, the amplitude and phase of the symbol is needed to distinguish a particular symbol.

In general, communication systems are full of trade-offs. The most important comes from a famous theorem developed by Claude Shannon which gives the maximum theoretical data bit-rate which can be communicated over a channel, known as the Shannon limit. This depends upon bandwidth, transmit power and receiver noise level, and gives the maximum data rate that can be sent over a channel, or link, with a given noise level and bandwidth.

## 17.5 Signal Upconversion

The process of mixing a baseband signal onto a carrier frequency is called upconversion, and is performed in the radio transmitter. In the radio receiver, the signal is brought back down to baseband in a process called downconversion. Traditionally, this up- and downconversion process was done using analog signals and analog circuits. Analog upconversion with real (not complex) signals is depicted in Figure 17.15.

To see why a mixer works, consider a simple baseband signal of 1 kHz tone and a carrier frequency of 600 kHz. Each can be represented as a sinusoid. A real signal, such as a baseband cosine, has both a positive and negative frequency component. At baseband, these overlay each other, so this is not obvious. But once upconverted, the two components can be readily seen both above and below the carrier frequency.

The equation for the upconversion mixer is:

$$\cos(\omega_{carrier}\ t) \times \cos(\omega_{signal}\ t)$$
$$= \frac{1}{2} \times [\cos((\omega_{carrier} + \omega_{signal}) \times t) + \cos((\omega_{carrier} - \omega_{signal}) \times t)]$$

The result is two tones, of half amplitude, located above and below the carrier frequency.

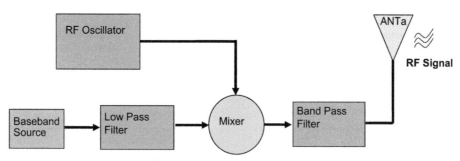

**Figure 17.15.** Analog upconversion.

## 17.6 Digital Upconversion

Alternately, this process can be done digitally. Let us assume that the information content, whether it is voice, music, or data, is in a sampled digital form. In fact, as we covered in the earlier chapter on modulation, this digital signal is often in a complex constellation form, such as QPSK or QAM, for example.

In order to transmit this information signal, at some point it must be converted to the analog domain. In the past, the conversion from digital to analog occurred when the signal was in baseband form, because the data converters could not handle higher frequencies. As the speeds and capabilities of analog-to-digital converters (ADC) and digital-to-analog converters (DAC) have improved, it has become possible to perform the up- and downconversions digitally, using a digital carrier frequency. The upconverted signal, which has much higher frequency content, can then be converted to analog form using a high-speed DAC.

Upconversion is accomplished by multiplying the complex baseband signal (with I and Q quadrature signals) with a complex exponential of frequency equal to the desired carrier frequency.

The complex carrier sinusoid can be generated using a lookup table, or implemented using any circuit capable of generating two sampled sinusoids offset by 90 degrees (Figure 11.3). In order to do this digitally, the sample rates of the baseband and carrier sinusoid signal must be equal. Since the carrier signal will usually be of much higher frequency than the baseband signal, the baseband signal will have to be interpolated, or upsampled, to match the sample frequency of the carrier signal. Then the mixing or upconversion process will result in the frequency spectrum shift depicted in Figure 17.16

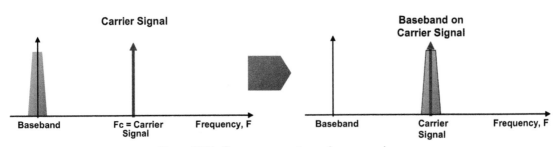

**Figure 17.16.** Frequency spectrum of upconversion.

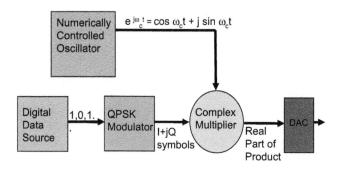

**Figure 17.17.** Digital upconversion.

For simplicity, assume the output of the modulator is a complex sinusoid of 1 kHz. The equation for this upconversion process is:

$$[\cos(\omega_{carrier}t) + j\sin(\omega_{carrier}\ t)] \times [\cos(2\pi \times 1000 \times t) + j\sin(2\pi \times 1000 \times t)]$$

$$= \frac{1}{2} \times [\cos((\omega_{carrier} - 2\pi \times 1000) \times t) + \cos((\omega_{carrier} + 2\pi \times 1000) \times t)]$$

$$- \frac{1}{2} \times [\cos((\omega_{carrier} - 2\pi \times 1000) \times t) - \cos((\omega_{carrier} + 2\pi \times 1000) \times\ t)]$$

$$+ j \times \frac{1}{2} \times [\sin((\omega_{carrier} + 2\pi \times 1000) \times t) + \sin\ ((\omega_{carrier} - 2\pi \times 1000) \times t)]$$

$$+ j \times \frac{1}{2} \times [\sin((\omega_{carrier} + 2\pi \times 1000) \times t) + \sin((-\omega_{carrier} + 2\pi \times 1000) \times t)]$$

This reduces to:

$$\cos\ ((\omega_{carrier} + 2\pi \times 1000) \times t) + j \times \sin\ ((\omega_{carrier} + 2\pi \times 1000) \times t)$$

The imaginary portion is discarded. It is not needed because at carrier frequencies both positive and negative baseband frequency components can be represented by the spectrum above and below the carrier frequency.

The final result at the output of the DAC is:

$$\cos\ ((\omega_{carrier} + 2\pi \times 1000) \times t)$$

Note there is only a signal frequency component above the carrier frequency. This is because the input is a complex sinusoid rotating in the positive (counter clockwise) direction – there is no negative frequency component in this baseband signal.

Normally, as the baseband signal will have both positive and negative frequency components. For example, the complex QPSK modulator trajectory can move both clockwise and counter-clockwise depending upon the input data sequence. When upconverted, the positive and negative baseband components will no longer overlay each other, but will be unfolded on either side of the carrier frequency. For example, a baseband signal with a frequency spectrum of 0 to 10 kHz will occupy a total of 20 kHz, with 10 kHz on either side of the carrier frequency. The baseband signal represents the positive and negative frequencies using quadrature form, which is why it is in the form of two dimensional I and Q signals, each with a frequency spectrum from 0 to 10 kHz.

A typical transmit chain in a cable video-distribution system would have many channels digitally modulated and placed upon different carrier frequencies. In this way, many video channels could be carried on the same coax or fiber. The NCO generates the complex exponetial signal. This process is shown in Figure 17.18.

In this implementation, there are N video channels with 6 MHz bandwidth being sampled at 16 MHz. The next step is to

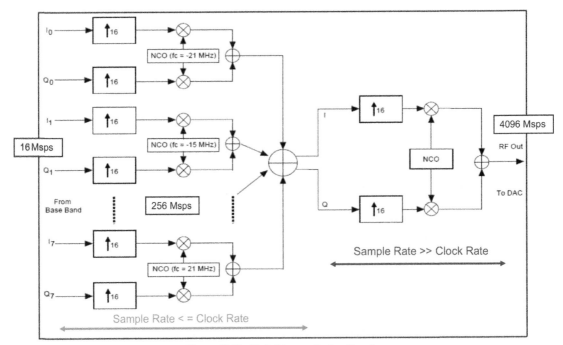

**Figure 17.18.** Digital transmit circuit path.

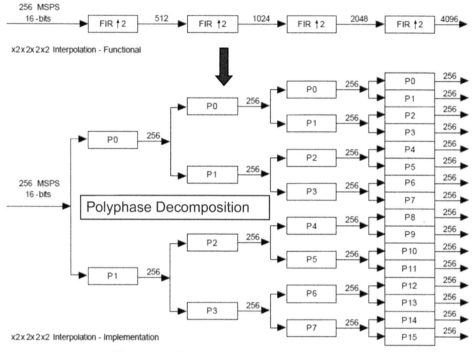

**Figure 17.19.** Polyphasing of high rate transmit paths.

pulse shape and interpolate, or up-sample each video signal by 16, to a sample rate of 256 MHz. Each of these baseband channels is placed upon a different carrier frequency, multiplying by the output of a numerically controlled oscillator (NCO), also running at 256 MHz, which is programmed to the desired carrier frequency.

The video carriers are all at separate intermediate frequencies, and so can be summed together. At this point, the composite signal can be digitally upconverted to RF frequency. Modern DACs can run at several GHz, and this design uses a DAC running at 4.096 GHz. This can allow for an RF signal in the multi-GHz range out of the DAC, which is filtered and amplified before being transmitted over the cable or broadcast over the air.

Most digital chips cannot clock at 4 GHz. Here, the portions of the circuit with data rates above 256 MHz are built using polyphasing techniques, which involve creating parallel paths in order to process a signal being sampled at rates above the digital clock rate. The sampling rate is stepped up from 256 MHz to 4096 MHz in four stages, each doubling the sample rate, for an overall interpolation of 16.

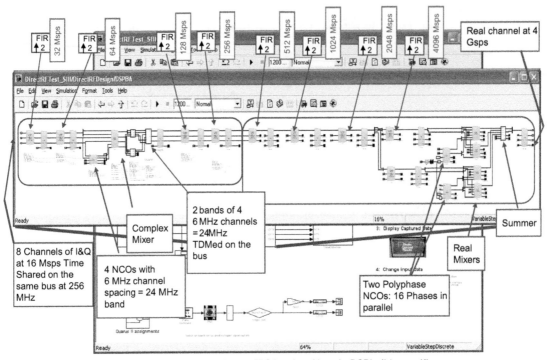

**Figure 17.20.** Implementation in FPGA using Altera's DSPbuilder toolflow.

The actual implementation of this design is shown in Figure 17.20. It is created using a toolflow known as DSPBuilder, from FPGA manufacturer Altera. The toolflow has the capability to automate the polyphasing, or paralleling of high data-rate circuits with a lower FPGA clock rate, in this case, 256 MHz.

# VIDEO OVER IP

## CHAPTER OUTLINE

18.1 Basics of Internet Protocol (IP)..................................................169
18.2 Encapsulation ............................................................................171
18.3 Video Streams............................................................................171
18.4 Transport Protocols ...................................................................172
18.5 IP Transport...............................................................................173
18.6 Video Over Internet Issues .........................................................175
18.7 Video Streaming.........................................................................176
18.8 Multicast Video ..........................................................................177
18.9 Video Conferencing.....................................................................178

Video over IP is becoming increasingly common. Ever larger numbers of people are switching from cable or satellite TV to IPTV, downloading programming over their internet connection. This can allow them to select what they want to watch, and avoid monthly cable or satellite fees. Websites such as YouTube provide both uploading and downloading of video content to millions of users. Even pornography is becoming an increasingly web-hosted video business.

## 18.1 Basics of Internet Protocol (IP)

Internet protocol is a communication system. Unlike the telephone system, it does not require a connection between the sender and receiver. Instead, the information is broken up into packets, and each packet finds its own path over the IP network from source to destination. These networks can be public such as the internet, or private such as a corporate network. The source, destination and every node in the network has an address, which is 32 bits, or eight hexadecimal bits. When expressed decimally, it is in the familiar form of xxx.xxx.xx.xx, or 10 decimal digits. The packets have two major components: the header and the data. The IP header is 20 bytes, and the data is a variable length up to 65615 bytes.

Digital Video Processing for Engineers. http://dx.doi.org/10.1016/B978-0-12-415760-6.00018-0

IP addresses can be constant, or temporary. This is best illustrated by example: an organization or corporation may have a unique, constant IP address, used to communicate over the public IP network. However, within the corporation there's a private IP network not intended to be accessible to outsiders, which may have hundreds or thousands of different computers or devices with IP addresses. Using dynamic host reconfiguration protocol (DHCP), a router within the corporation can assign temporary IP addresses to these nodes on the private network. Certain ranges of IP addresses are designated for use by DHCP. These addresses are assigned whenever a computer tries to connect to the private network. The dynamically assigned addresses need only to be unique within the private network — other private networks can use the same range of addresses within their own private network. A function in the router bridging the private network to the external network, called network address translation (NAT), is used to translate between the address space of the public internet and the DHCP-assigned addresses within the private network.

Routers are the key components that allow anyone to transmit data to anyone else over the internet. Routers are distributed throughout the IP networks — they examine the headers, and, using the destination address, forward the packets towards their destinations. Since the packet may pass though many routers, each router must decide how best to forward the packet to get it closer to its destination. This is done by maintaining large routing tables, and monitoring the status of various network connections for traffic levels, and sometimes by determining priority for a given packet.

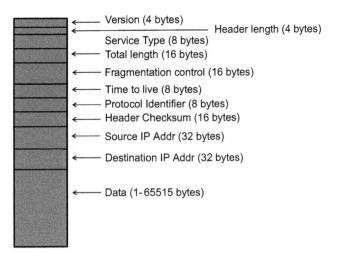

**Figure 18.1.** IP packet formatting.

## 18.2 Encapsulation

The video data may be continuously streaming or form a single large file. However it is formatted, it must be broken up into chunks to fit into IP packets, a process called encapsulation. A simple method could use the largest possible data size for each IP packet. However this has several disadvantages. Using very large IP packets increases latency, because the packet cannot be sent until enough data is made available by the source. If a packet gets corrupted over the network, a large and noticeable amount of data is lost. Also, large packets may be fragmented, meaning they may be broken into multiple packets during transmission so that they can be used by some networks − for example, standard Ethernet only allows transmission of up to 1500 bytes in packet.

Very short packets, on the other hand, are inefficient. Since the packet header is a constant size, it uses a larger portion of the total packet size. More packets also put more load on the routers in the network to process and route, as each packet is treated individually. Sensitivity to latency as well as the packet error rates on a given network can be used to determine packet data lengths.

## 18.3 Video Streams

Compressed video uses several types of streams for transport over a network. The simplest is the elementary stream, which contains just the compressed data output from the video encoder and does not contain audio or synchronization data.

A program stream contains several elementary streams, for video, audio or data. It contains everything needed for a given program to be presented. The data might be for on-screen text message overlay, or it might be used for production and recording functions. Time stamps are added to each of the elementary streams to synchronize them. These enhanced elementary streams are called packetized elementary streams (PES) because the elementary stream has been broken into packets, each associated with a different timestamp. The PES should not be confused with IP packets. PES is associated with how video data is packaged to maintain synchronization in a specific protocol used for compressed video, and has nothing to do with what method or protocols are used to move data, of which the internet is just one of several methods.

There are two time stamps: one is known as the presentation timestamp, indicating when each video packet should be displayed in the video; the other is the decode timestamp, indicating the order that it should be processed by the decoder. As discussed in previous chapters, the order of decode is different from the order of display in a typical GOP, with a mixture of I-, P- and B-frames.

Program streams can contain multiple video displays, supporting picture in picture, or multiple video sources — such as an anchor person on one part of the screen and a remote video feed in another. Another example is sports coverage, where multiple angles might be shown simultaneously. Program streams are usually used when little or no data loss is expected in the transmission, and typically use long packets. Applications of program streams include DVD players or within a production studio or trailer.

The transport stream is often used when the video is transmitted over long distances, over different types of multi-user networks. This could be satellite links, broadcast terrestrial links or video over IP. Transport streams feature error correction schemes, such as Reed Solomon. To facilitate error correction, fixed-length packets of 188 bytes are used. Accounting for additional bytes in error correction, this can increase to 204 or 208 bytes. Transport streams can contain multiple elementary streams, and each is identified with a packet identifier (PID). PIDs are used to differentiate between the video, audio and data elementary streams used to make up a complete program. To keep track of the PIDs, two further structures are used. One is the program association table (PAT), which is sent to provide the index of all the programs in the transport stream (this is used if PID $= 0$). The other structure is multiple program map tables (PAT), one for each program, giving the PID numbers for the video stream(s), audio stream(s) and data stream(s) in that program.

The transport-stream packets are much smaller than the allowable data-packet lengths used in IP. Even restricting the IP packet length to 1500 bytes to prevent fragmentation on Ethernet, there can be seven of the 208 byte long transport stream packets encapsulated.

## 18.4 Transport Protocols

Video streams are sent over IP networks. IP networks have their own protocols, which are independent of what type of data or application the IP network is being used for. These protocols are not specific to video over IP, but are used for nearly all internet applications.

The most common protocol used is transmission control protocol (TCP), also commonly referred to as transmission control protocol over internet protocol (TCP/IP). TCP establishes a reliable connection between source and destination using defined handshaking schemes. It keeps track of all IP packets by assigning a sequence identifier. TCP can detect if a packet is missing at the destination, and keep packets in order, even if some packets get delayed across the network. If a packet is corrupted, or fails to arrive, the destination can request a retransmission from the source.

This is extremely valuable for many types of service using the internet. For example, when sending an email with a file attachment, every byte must arrive and be reassembled in the correct sequence in order to be useful. Commerce over the internet demands error-free communication.

However, this protocol is ill-suited for some types of service, including streaming video and audio: if an IP packet is corrupted or substantially delayed, it makes little sense to request a retransmission, as the data is needed in a timely manner or not at all. Requesting retransmissions needlessly uses up more network resources and bandwidth. Also, TCP has provisions to reduce data rates if there are too many corrupted packets-known as flow control-and reducing the data rate can be very disruptive to a streaming application. For example, in a voice over IP phone call, having a few clicks due to lost data is preferable to pauses of silence while waiting for all the audio data to arrive correctly. TCP also incurs more latency or delay due to its error-free connection features.

A simpler alternative protocol is user datagram protocol (UDP). There is no handshaking between source and destination; the data is just sent. There are no attempts at tracking missing IP packets or retransmitting, or flow control. It is low latency, simple and low overhead. It is more suited for streaming data applications. If the IP packets contain data that has its own error correction built in, like transmit streams, then corrupted data may be corrected.

An extension of UDP is real-time protocol (RTP). RTP is specifically designed for real-time data-streaming applications which cannot tolerate interruption in data flow, and need minimum latency. RTP does provide some additional features compared to UDP: a time-stamping feature to allow multiple streams from a given source to be synchronized, such as video and audio; multi-casting support, so one source can send the same data to many destinations simultaneously; and packet sequencing, so lost packets can be detected, which can allow a video decoder, for example, to use previous or nearby video data as a best guess for the lost data. Note the actual protocol used with RTP is real-time streaming protocol (RTSP).

## 18.5 IP Transport

Video over IP can be physically transmitted using many methods, some of which are listed below.

Ethernet is familiar to most of us, as it's commonly used within buildings or campuses and is known as a local area network (LAN). Ethernet uses another address, appended to the packet, known as the media access control (MAC) address — this is a permanent address assigned by the manufacturer of that equipment. The MAC address is a 12-hexadecimal digit address xx:xx:xx:xx:xx:xx. MAC address ranges are assigned by an international agency to manufacturers of products containing Ethernet ports. For transport within an Ethernet LAN the MAC address, appended to the IP packet, is used. LANs are built up of Ethernet equipment (computers for example), Ethernet hubs, Ethernet switches and Ethernet bridges. Hubs simply act as repeaters as any packet coming in on one port will be sent out on all the other ports. Switches examine MAC addresses, and only forward packets on ports that connect to the MAC addresses for those packets. Bridges connect different LANs together. These could be all Ethernet based LANs, or could be wireless LANs using 802.11 based wireless technology. Common Ethernet speeds are 100 BT (100 Mbps) or gigabit Ethernet (1 Gbps). Faster Ethernet speeds, such as 10 Gbps, are possible on specially designed backplanes or over fiber interfaces.

There is a wider range of technologies used for wide area networking (WAN), which, as the name implies, is over distances ranging from a few miles to thousands of miles. These are usually owned by a carrier, and used as shared resources (the internet for example), or connections can be leased and used privately (connections between corporate offices in different locations, for example). However, IP is still used as the protocol over the different transport services and technologies.

Synchronous Optical Network (SONET) and Synchronous Digital Hierarchy (SDH) are two popular standards used for transmission over fibers, and form the basis of most long-distance telecommunications. The speeds used can be above 10 Gbps. Asynchronous Transfer Mode (ATM) is a protocol used in these networks. ATM operates by allocating a specific amount of bandwidth to a given connection using virtual circuits. It allows for much finer control over the data bandwidth allocated for a given connection or user. However, due to the "guaranteed" bandwidth allocation, ATM tends to be an expensive way to communicate when bandwidth requirements are dynamic.

Fortunately, IP can be layered over ATM, and the user will not even be aware of the ATM protocol running underneath. Cable and digital subscriber lines (DSL) are used for intermediate length connections, typically from IP service providers to homes and small businesses. This is often referred to as "last mile". Rather than using fiber, the physical connections are made with coaxial

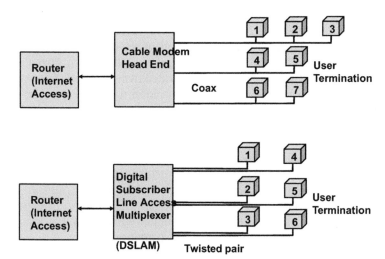

**Figure 18.2.** Aggregated access to internet.

or twist pair wiring. These lower-speed connections are aggregated into a single higher-speed connection by the service provider, at what is called a central office (though this function could just be an equipment cabinet).

The common technology used across WANs, last mile and LANs is IP and its protocols such as TCP/IP, UDP, and RTP. In particular, compressed video over IP can be streamed across networks made up of all these technologies.

## 18.6 Video Over Internet Issues

The internet is a remarkable network for all sorts of communication, but it does have serious limitations for video transmission. The available raw bandwidth is large, but as there is a high variability in number of users and demand from those users, so available bandwidth for a given user is never guaranteed. Packet losses also occur, perhaps as much as 1%. More problematic is the amount of jitter, or variation on the latency of transmission from packet to packet. A packet that arrives late is the same as a packet that never arrives in a real-time system.

These issues are slightly mitigated by the latest video compression algorithms, which reduce video bandwidth, particularly on low-resolution video delivery. Many program sources of video are available on the internet.

Dedicated non-internet transmission is needed for high-definition video with good quality. This is why the great majority of consumers subscribe to a cable or satellite service, with high quality, dedicated transmission networks of video content.

Movie downloads can be delivered efficiently through the internet: companies like Netflix are moving away from DVDs by mail, towards internet downloads of compressed movie content. One advantage is that the movie can be downloaded in non-real-time, mitigating the issues of streaming video over IP. This is known as download and play. Progressive download and play is some-where in the middle: the video is broken up into segments, and played as soon as one segment is finished downloading. As long as the next segment can finish downloading prior to the completion of the previous segments play time, there is no interruption.

There are many video applications besides movies. Download and play of short YouTube video clips (often low resolution), marketing webcasts, investor and analyst briefing or private corporate broadcast of executive's speeches are all applications that can be well suited to video over IP transport.

## 18.7 Video Streaming

Video streaming requires real-time performance of the IP network to deliver the compressed video content. Usually RTP is used over the IP network.

The video source can be a video streaming server, a computer or a webcam. The video server typically has a large storage capability to house the large amounts of video to be broadcast on demand. Alternatively, the video may be live (or almost live), where it is being recorded by a camera, formatted, compressed and then sent out for viewing across the internet (perhaps a webcam). The formatting of the video can be of multiple forms, different video players being supported with one or more of these formats.

The software to do the video formatting and playing is avail-able from several companies, and each has its proprietary methods. Some of the familiar names are: Windows Media Player by Microsoft; QuickTime by Apple; RealPlayer® by RealNetworks and Adobe.

Newer video-streaming standards such as HTTP Live Streaming from Apple have been developed to support video streaming to iPhones and other smart mobile devices. This standard uses HTTP (Hypertext Transfer Protocol) IP technology as opposed to RTSP, which can allow it to bypass many firewalls in IP networks. Microsoft offers Smooth Streaming, which also dispenses with RTSP in favor of HTTP IP technology.

HTTP is not designed for video streaming, but it has been found to be very efficient. It was originally designed for file transfer, and not to maintain a persistent connection. More recently, a keep-

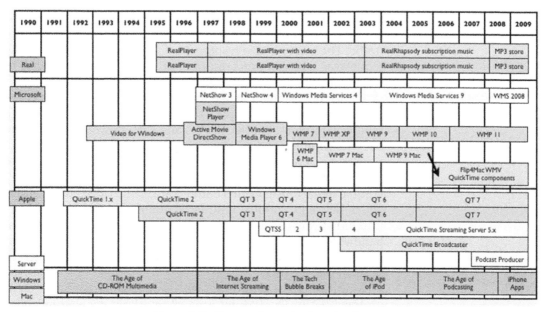

**Figure 18.3.** IP streaming technology development.

alive mechanism was introduced, where a connection could be reused for more than one request. Using a persistent connection reduces the latency, because the TCP connection does not need to be to re-negotiated after the first request has been sent.

The progression of products enabling streaming video and audio is summarized in Figure 18.3.

Most of the discussion to this point has been about unicasting, or sending video from one source to one destination. However, many applications are multicast, such as live events, broadcast style IP TV or security system and traffic cameras.

## 18.8 Multicast Video

Multicast could be accomplished by running many parallel unicasts, assuming the video server has enough aggregate bandwidth to support many video streams in parallel. Even if the bandwidth did exist, it would be very inefficient. Instead, multitasking is primarily supported in the router. The router is required to recognize packets being multicast, replicate them, and send to multiple destinations or addresses. This is not to be confused with IP broadcasting, where a single packet is sent to all devices on the local network.

Multicasting is generally not available on the public internet but it can be enabled on private or corporate networks. It places a large burden on the router however, and this should be considered, given the router is generally supporting a lot of other traffic. The router is also responsible for detecting and processing requests to add new ports or drop ports as users either request to watch, or drop off, the program.

Unlike unicasting, the user has no control over the delivery of a multi-cast. When connecting they will join at whatever point the multicast program happens to be, and each viewer sees the same content simultaneously, similar to broadcast television. These restrictions have to be set against the significant advantages of network traffic. Just as the video bandwidth from the video server is equal to the unicast video bandwidth, so the video bandwidth is equal between various routers. If there are multiple viewers watching from a downstream router, then only the unicast bandwidth is required. Only if viewers are connected to different routers does the video stream need to be replicated at different parts of the network.

Multicast addressing is set up using session announcement protocol (SAP). The SAP informs all multicast-enabled receivers of the programs being multicast broadcast on the network. The details of either connecting or disconnecting from a multicast are covered in the Internet Group Management Protocol (IGMP). Using IGMP, the routers need to keep track of all users in their downstream path, to know whether or not the multicast program is to continue to be broadcast on a given port. If a new user requests to join, the appropriate router must replicate the program in that portion of the network if it is not already being broadcast. The routers will also broadcast the SAP messages.

## 18.9 Video Conferencing

Video conferencing that feels like everyone is in the same room has been a business goal for a long time. Video conferencing can be done over IP, using private networks, but latency must be carefully controlled so the interactions and conversations between people are not delayed.

The H.320 standard defines video conferencing over switched telephone lines (ISDN), and is used in much early corporate video-conferencing equipment, usually in dedicated conference rooms at different sites within a company. H.323 evolved from H.320, allowing video (and audio) conferencing to be packet based. It is a full-featured system, with considerable maturity. A number of vendors offer systems based upon H.323, and generally equipment from different vendors will interoperate. Most systems support only audio, as audio or voice conferencing is far more ubiquitous than video conferencing. The video conferencing options of H.323 feature video compression using H.264.

Another standard known as session initiation protocol (SIP) has become very popular for packet-based audio and video conferencing. SIP focuses on the messaging needed to set up, maintain and tear-down joint sessions between multiple users. The actual transport of the audio, video and data is over RTP. SIP also has additional features, such as instant messaging, and is generally more flexible than H.323.

Despite efforts to encourage the adoption of video conferencing as a primary means of holding long distance meetings, audio conferencing remains far more prevalent. However, live sharing of joint data or presentations in conference calls using technology like WebEx™ or Office Communicator has become commonplace in the business world, leveraging the capabilities of SIP.

# SEGMENTATION AND FOCUS

**Steve Fielding, Base2Designs**

## CHAPTER OUTLINE

**19.1 Measuring Focus   182**
    19.1.1 Gradient Techniques   182
    19.1.2 Variance Techniques   184
**19.2 Segmentation   185**
    19.2.1 Template Matching   185
    19.2.2 FPGA Implementation   190

The majority of image-processing algorithms require a properly focused image for best results. For some applications this may not be difficult to achieve because the camera capturing the image will have a large depth of field: objects at a wide range of distances from the camera will all appear in focus without having to adjust the focus of the camera. However, for cameras that have narrow depth of field, the image will not always be in focus, and there needs to be some way of assigning a focus score to an image — a measure of focus "goodness". The focus score serves two purposes:

Out-of-focus images with low focus score can be rejected. This improves the results from image processing algorithms that require focused images and reduces the processing load on these algorithms.

The focus score can be used in the feedback control of a camera autofocus mechanism.

Because focus-score assessment is at the front end of any image-processing system, it has a large impact on system performance. For example, imagine a video stream that contains focused and unfocused images, and feeding these to an image processor that takes one second to process each video frame. The system would be generating a lot of bad results, and also taking a long time to generate a good result. In fact, if the image processor is randomly sampling the incoming video frames (i.e. there is no video storage), it may never produce a good result.

Digital Video Processing for Engineers. http://dx.doi.org/10.1016/B978-0-12-415760-6.00019-2

## 19.1 Measuring Focus

We all recognize focused and unfocused images when we see them. Unfocused images are characterized by blurry and difficult to identify objects, whereas focused images are characterized by well-defined, easily identifiable objects. Let's take a look at what we mean by objects appearing blurry: an object that appears blurry has edges that are "fuzzy", and the transition from object to background is gradual. Compare this to a focused image, where the edges of objects are sharp, and the transition from object to background is immediate. The effect of focus on edge sharpness can be seen in Figure 19.1. We can effectively generate a focus score for an image or an object within an image by measuring the sharpness of these edges.

### 19.1.1 Gradient Techniques

Gradient techniques measure the difference between adjacent pixel grayscale values, and use this as a measure of rate of change of grayscale, or gradient of the gray levels. Focused images exhibit large rates of change in pixel gray level, whereas in an unfocused image the gray level value's rate of change is lower. Thus we can measure the absolute gradient of every pixel, sum all the gradients and use this as a focus score.

$$F = \sum_{y=0 \text{ to } M-1} \sum_{x=0 \text{ to } N-1} | i_{(x+1,y)} - i_{(x,y)} |$$

Where:
F = Focus score
N = Number of pixels in image row
M = Number of pixels in image column
i(x,y) = Image gray level intensity of pixel (x,y)

**Figure 19.1.** Example of focused and unfocused images.

$$| i_{(x+1, y)} - i_{(x,y)} | > = T$$

T = Threshold gray level

Let's calculate the focus score in the horizontal direction for the two images shown in Figure 19.2.

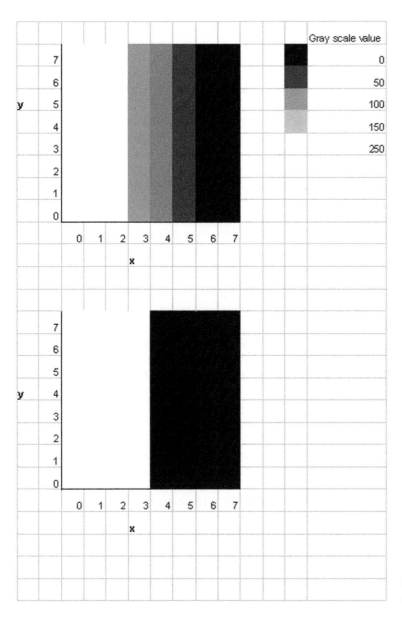

**Figure 19.2.** Example of focused and unfocused vertical edge.

We'll set T $= 100$ and we will firstly calculate the focus score for the unfocused image, with the blurry edge:

F $= (250 - 150) \times 8 = 800$

Note that there is only one difference that is greater than or equal to T. Since every row is the same, we multiply this one diference by the number of rows.

Next we will calculate the focus score for the focused image, with the sharp edge:

F $= (250\text{-}0) \times 8 = 2000$

From this calculation we see the focused image has a higher score. Thus for multiple images of the same scene we have a method that can identify images with the best focus.

This rate of change also translates to higher frequencies in the spatial domain, so applying a high-pass filter and then measuring the power of the filtered image gives an indication of the total power of the high-frequency components in the image and thus the focus. Noise within an image can also have high-frequency content: depending on the system, it may be necessary to use a band-pass filter that rejects some of these high frequencies.

## 19.1.2 Variance Techniques

Since the pixel gray values of edges in a focused image transition rapidly, such images exhibit greater variance in pixel values. We can measure this by calculating the difference between each pixel and the mean value of all the pixels. We square the difference to amplify larger differences, and remove negative values:

F $= (1\ /\ \text{MN}) \sum_{y\,=\,0\text{ to }M\,-\,1} \sum_{x\,=\,0\text{ to }N\,-\,1} (i_{(x,y)} - \mu)^2$

Where:

$\mu = $ Mean of all pixels.

Let's apply this variance measure to our two images in Figure 19.2. First we we'll calculate the mean for the unfocused image:

$\mu = (250 + 250 + 250 + 150 + 100 + 50 + 0 + 0) \times 8 /$
$64 = 131.25$

Next we calculate the variance:

F $= (1\ /\ 8 \times 8) \times (((250 - 131.25)^2 + (250 - 131.25)^2 +$
$(250 - 131.25)^{\,2} + (150 - 131.25)^2 +$
$(100 - 131.25)^2 + (50 - 131.25)^2 + (0 - 131.25)^2 +$
$(0 - 131.25)^{2)} \times 8)$
F $= 10{,}586$

Similarly for the focused image:

$\mu = (250 + 250 + 250 + 250 + 0 + 0 + 0 + 0) \times 8 / 64 = 125$

$F = (1 / 8 \times 8) \times (((250 - 125)^2 + (250 - 125)^2 + (250 - 125)^2 + (150 - 125)^2 + (0 - 125)^2 + (0 - 125)^2 + (0 - 125)^2 + (0 - 125)^2) \times 8)$

$F = 15{,}625$

Here we see that the focused image has a higher variance than the unfocused image.

One of the variations on this algorithm divides the final result by the mean pixel value, thus normalizing the result, and compensating for variations in image brightness.

# 19.2 Segmentation

Until now we have considered the entire image to be in the same level of focus. This may not be the case. For example, images of tiny micro-electro-mechanical-systems (MEMS) may have regions of the image that are in different levels of focus, thus we have to choose the region of the image that we want to be correctly focused. Sometimes this can be as simple as choosing a region in the middle of the image, but it is not always this easy, and there are situations where the region of interest must be located and its focus score calculated. We would need to extract a segment from the image. There are techniques to find the edges in an image, and from this information identify the edges of interest. These techniques are beyond the scope of this chapter — what we will consider here is the simpler case where we know the shape of the region that we want to locate.

## 19.2.1 Template Matching

The segmentation algorithm uses expanding, shifting shape templates to locate regions within the image. For example, if the target region is a circle, then pixel image data is collected from circular regions of varying radius and position. Summing pixel data in these circular regions gives a gray level score for the region and comparing gray values between circular regions determines the region boundary.

Let's take a look at the example in Figure 19.3. We are going to locate a square region of $4 \times 4$ pixels, located within a $12 \times 12$ image. Since we are looking for a square region, we will use shifting, expanding square templates and match these against the image. Furthermore, since the square region we are trying to locate has uniform gray values, we will only add the values of the pixels on the perimeter of the squares, and ignore the pixel values inside the squares. We will start with a square template of size

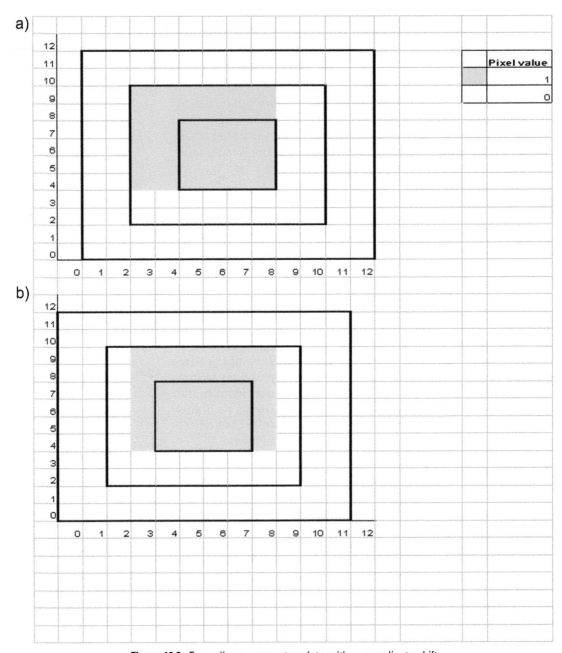

**Figure 19.3.** Expanding a square template with × coordinate shift.

four at coordinates (6, 6) Figure 19.3(a). First we will add or integrate the values around the square, and normalize by dividing by the total number of pixels:

$S(4) = (12 \times 1) / 12 = 1$

Where:

$S(n)$ = Sum of all the pixel values around a square with sides of length n.

Similarly for template squares of size 8 and 12:

$S(8) = ((11 \times 1) + (17 \times 0)) / 28 = 0.36$
$S(12) = (44 \times 0) / 44 = 0$

Now we calculate the maximum gradient of the integral sums:

$G(4,8) = | 1 - 0.36 | = 0.64$
$G(8,12) = | 0.36 - 0 | = 0.36$

Where:

$G(n,m)$ = The gradient between squares of with sides length n, and m.

Thus for expanding squares with origin (6, 6) the maximum gradient is 0.64.

Now we will shift the origin to (5, 6) as shown in Figure 19.3(b). First we calculate the integrals:

$S(4) = (12 \times 1) / 12 = 1$
$S(8) = ((6 \times 1) + (22 \times 0)) / 28 = 0.21$
$S(12) = (44 \times 0) / 44 = 0$

And now the gradients:

$G(4, 8) = | 1 - 0.21 | = 0.79$
$G(8,12) = | 0.21 - 0 | = 0.21$

Notice that the peak gradient is higher for the second position, where the expanding circles are centered at (5, 6). This tells us that (5, 6) is more likely to be the origin of the square than (6, 6). That is $x = 5$ is the best candidate for the x coordinate of the origin.

We can perform the same shifting operation in the y direction, and thus find the optimal y coordinate as illustrated in Figure 19.4. If we calculate the gradients we get a similar result to that for the x shifting, and in this case we find that (6, 7) is a better candidate than (6, 6) and thus $y = 7$ is the best candidate for the y coordinate of the origin. Our best estimate so far for the origin of the target square is $x = 5$, $y = 7$ or (5, 7). If we check again we will see that (5, 7) is indeed the origin of the square. In a real example we would not be so lucky, and we would need to repeat the operation a number of times. In each iteration, we would use the

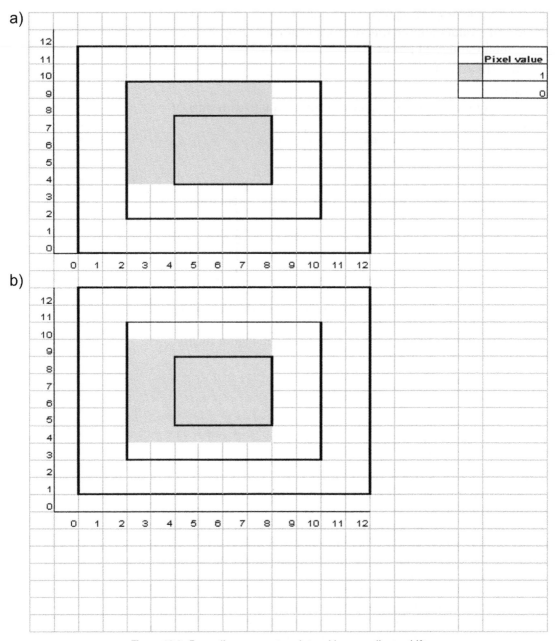

**Figure 19.4.** Expanding square template with y coordinate shift.

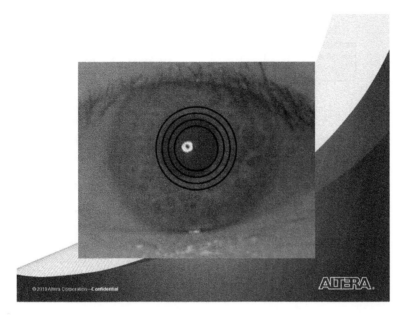

**Figure 19.5.** Expanding circle template applied to image of eye.

newly calculated origin and a reduced step size for the x and y shifts.

We also see that the gradient is larger between the squares of side length four and eight pixels than it is between the squares of side length 8 and 12 pixels. This tells us which template squares are more likely to border the target square. So, in each iteration of the algorithm, as well as reducing the x and y step sizes, we would also use the new best-candidate template-square size, and reduce the step size of the expansion.

In the previous example we used a contour integral to determine the average gray level at each template position. This works well if the region has uniform gray values but, if this is not the case, it may be necessary to perform an area integral instead of a contour integral. This has the effect of averaging the gray values within the template area, and thus reducing the effect of image details.

Once the target region has been located within the image, then one of the focus algorithms previously described can be applied to the region, and a focus score calculated.

Figure 19.5 shows the use of expanding circles to locate the pupil within an image of the eye. The process is similar to the one described for expanding and shifting square templates, but in this case the template is a circle.

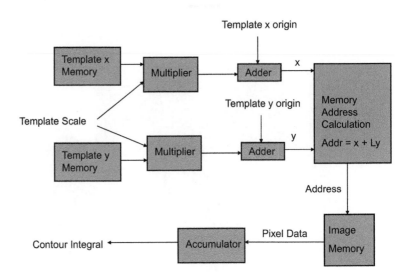

**Figure 19.6.** Block diagram of a contour integral engine.

## 19.2.2 FPGA Implementation

Figure 19.6 shows a block diagram of a system designed to perform a contour integral. The base template is stored as x and y coordinates in two separate memories. Template expansion is performed by multiplying the template coordinates by a scaling factor, and the scaled template coordinates are fed to an adder. Here the x and y offset coordinates are added to generate an (x, y) coordinate reference within the image frame. This (x, y) reference must be manipulated to create a linear address within the image frame store. Assuming image rows are stored in contiguous memory blocks, this would involve multiplying y by the image line width and adding x. The generated address can be used to retrieve the pixel data from memory, and the data fed to an accumulator which will sum all the pixel data for the contour defined by the scaled and shifted template.

By pipelining the operations, it is possible for an FPGA to perform all the operations in parallel, so that memory-access speed becomes the performance limitation. Depending on the image size, it may be possible to store the image in internal FPGA RAM. If this is possible then not only can the memory operate at higher speed, but the dual-port feature of the internal RAM can be used so that two of these contour integral engines can be executed in parallel.

# 20

# MEMORY CONSIDERATIONS WHEN BUILDING A VIDEO PROCESSING DESIGN

## CHAPTER OUTLINE

20.1 The Frame Buffer    191
20.2 Calculating External Memory Bandwidth Required    194
20.3 Calculating On-Chip Memory    199
20.4 Conclusion    200

Video processing — especially HD video processing — is significantly compute and memory intensive. When building an HD video-processing signal chain in hardware, whether in ASIC or FPGA, you must be aware of both the computation resources and the internal and external memory resources that will be needed.

This chapter will use simple examples of designs to explain how to calculate memory resources. First we will look at external memory bandwidth since in any HD video processing system the pixels required for calculation cannot be stored on-chip. External DDR memory is also a very important consideration. To ascertain this bandwidth we will look at the frame-buffer function and the motion-adaptive deinterlacing function.

We will then assess internal on-chip memory requirements — especially for functions like the video scaler which processes lines of video within a given frame.

## 20.1 The Frame Buffer

This function is widely used in a range of video-processing signal chains. Frame buffers do exactly what the name suggests — it buffers video frames, not just lines, in an external DDR memory. Buffering is frequently required to match the data rates and thus reduce video burstiness. Frame buffers take in a frame of

Digital Video Processing for Engineers. http://dx.doi.org/10.1016/B978-0-12-415760-6.00020-9

video — line by line — and transfer it to an external memory. Then the buffer will read either that frame, or any other from the DDR memory, into the internal on-chip memory. It may not transfer the entire frame, usually transferring a few lines of video which are required for processing.

To implement this functionality a buffer needs to have a writer block and a reader block. It is built with a writer which stores input pixels in memory, and a reader which retrieves video frames from the memory and outputs them. See Figure 20.1.

Figure 20.1 shows a simple implementation of two types of frame buffer — a double frame buffer and a triple frame buffer. As the name suggests, a double frame buffer stores two frames in the external DDR memory, while a triple frame buffer stores three frames in memory.

Let's look first at the double frame buffer. When double buffering is in use, two frame buffers are used in external RAM. At any time, one buffer is used by the writer component to store input pixels, while the second buffer is used by the reader component that reads pixels from the memory. When both the writer and reader finish processing a frame, the buffers are exchanged. Double buffering is often used when the frame rate is the same at the input and at the output sides, but the pixel rate is highly irregular at one or both sides. For example, if a function is putting out one pixel on each clock edge, but the next function needs nine pixels in order to do its work, you would insert a frame buffer between the two functions. Double buffering is useful in solving throughput issues in the data path and is often used when a frame has to be received or sent in a short period of time compared with the overall frame rate.

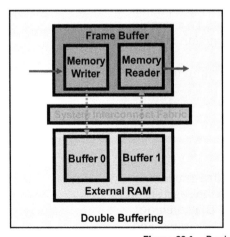 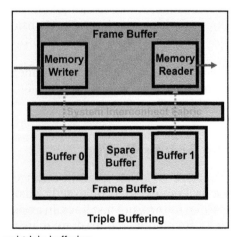

**Figure 20.1.** Double and triple buffering.

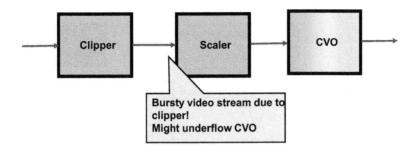

**Figure 20.2.** Clipping a video frame before it is passed it on to a video scaler.

When frame dropping and frame repeating are not allowed, the frame buffer needs to provide a double-buffering function. However, if frame dropping and/or frame repeating are allowed, the frame buffer provides a triple-buffering function and can be used to perform simple frame-rate conversion.

When triple buffering is in use, three frame buffers are used in external RAM. As in the case of double buffering, the reader and writer components are always locking one buffer to respectively store input pixels to memory and read output pixels from memory. The third frame buffer is a spare buffer that allows the input and output sides to swap buffers asynchronously. Triple buffering allows simple frame-rate conversion to be performed when the input and output are pushing and pulling frames at different rates. Also, by further controlling the dropping or repeating behavior, the input and output can be kept synchronized.

Let's look at some examples where you may need to insert a frame buffer. Figure 20.1 shows an example where we are clipping a video frame before passing it on to a video scaler. This is commonly used to zoom into a portion of an image.

First we clip a corner of the video frame. The rest of the pixels are now not valid — only the clipped pixels go on to the scaler. While the frame rate (frames/sec) is the same, the pixel rates that are input to the clipper and to the scaler are different. If the clipper output is passed directly on to the scaler, we will have periods of valid pixel data and periods of not valid pixel data — we effectively have a bursty video stream after the clipper.

The bursty video stream would cause the scaler to stall from time to time — and this will impact downstream processing. To mitigate that effect, a double-buffering frame buffer should be inserted between the clipper and the scaler.

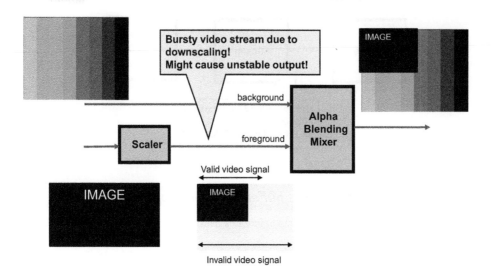

*Solution: Insert frame buffer between scaler and Alpha Blending Mixer*

**Figure 20.3.** A down-scaler output connected an alpha blender.

Another example is shown in Figure 20.3. This shows what happens when a down-scaler output is connected to a function, such as an alpha blender, which blends two or more video streams. A down-scaler also produces fewer pixels than it takes in. Therefore the output of this function is bursty – valid pixels followed by not valid data. A down-scaler is also called a decimation filter and as a rule these filters will produce bursty video data.

As before, we should insert a double-buffering frame buffer between the scaler and the alpha blending mixer to minimize the chance of starving all the downstream modules and creating unstable output at the alpha blending mixer.

As a matter of fact, 2-D FIR filters, 2-D Median filters and deinterlacers (with Bob algorithm) all produce bursty video streams. Depending on the application requirement, a double-buffering frame buffer might need to be connected after each of these functions.

## 20.2 Calculating External Memory Bandwidth Required

In video-signal chains the two functions that require lots of external memory transactions are the frame buffer and the deinterlacer. The frame-buffer memory transactions are described

above. The deinterlacer, especially the motion-adaptive deinterlacer, requires external memory transactions because it com-
pares fields and determines if there is motion or not and that's
where the external memory transactions come into play.

In this chapter we will calculate the memory bandwidth for
a hypothetical signal chain that includes a frame buffer and
a deinterlacer — so that you can appreciate the staggering amount
of memory bandwidth required in HD video processing.

Figure 20.4 shows an example video design. We will focus on
the two deinterlacers and the two frame buffers, and for each of
these components we will calculate the worst-case memory
bandwidth.

A deinterlacer has to read four fields (or for calculation
purposes two frames) and calculate the motion value (vector) for
each pixel. This value is then compared with the previous value of
the motion vector. So it has to read in the motion vector and then
write out the final motion vector.

As shown in Figure 20.5, a motion-adaptive deinterlacer
requires five (master) accesses to the DDR memory:

One field write (@input rate).

Two field reads (@output rate) — four fields in two accesses.

One motion vector write.

One motion vector read.

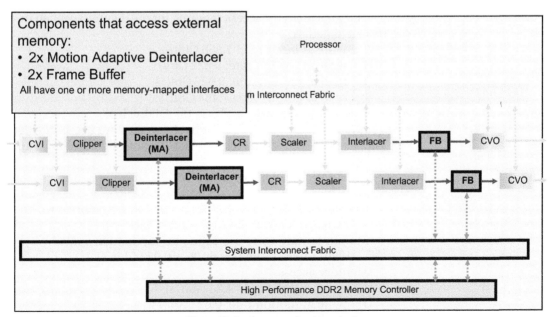

**Figure 20.4.** An example video design.

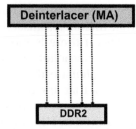

**Figure 20.5.** A motion-adaptive deinterlacer requires five (master) accesses to the DDR memory.

So to calculate the memory bandwidth demanded by this deinterlacer, we will first assume that this is a simple 4:4:4 video with 10 bits per color plane for each pixel. This means that each pixel requires 30 bits to be represented. The input format to the deinterlacer is fields and the output format is frames.

Input format: 1080i, 60 fields/sec, 10-bit color

- $1920 \times 1080 \times 30\text{bits} \times 60/2 = 1.866\,\text{Gbit/s}$

Output format: 1080p, 60 frames/sec, 10-bit color

- $1920 \times 1080 \times 30\text{bits} \times 60 = 3.732\text{Gbit/s}$

Let's also assume that the motion vector calculated is represented as a 10-bit value. So there will be one motion vector read and one motion vector write.

Motion format: only use 10 bits for the motion values:

- $1920 \times 1080 \times 10\text{ bits} \times 60/2 = 0.622\text{ Gbit/s}$

The total memory access required for the deinterlacer can thus be calculated as:

$1 \times$ write at input rate: 1.866 Gbit/s.

$1 \times$ write at motion rate: 0.622 Gbit/s.

$1 \times$ read at motion rate: 0.622 Gbit/s.

$2 \times$ read at output rate: 7.464 Gbit/s.

Total: 10.574 Gbit/s ← this is for 4:4:4 video.

Just as an exercise, let's see what happens if we use 4:2:2 video. The pixels are represented by 20 bits, however the motion vectors are still at 10 bits.

Input format: 1080i, 60 fields/sec, 10-bit color

- $1920 \times 1080 \times 20\text{bits} \times 60/2 = 1.24\text{ Gbit/s}$

Output format: 1080p, 60 frames/sec, 10-bit color

- $1920 \times 1080 \times 20\text{bits} \times 60 = 2.48\text{ Gbit/s}$

Motion format: Only use 10bits for the motion values

- $1920 \times 1080 \times 10\text{bits} \times 60/2 = 0.622\text{ Gbit/s}$

Memory access:

$1 \times$ write at input rate: 1.24 Gbit/s.

$1 \times$ write at motion rate: 0.622 Gbit/s.

$1 \times$ read at motion rate: 0.622 Gbit/s.

2 × read at output rate: 4.96 Gbit/s.

Total: 7.44 Gbit/s ← this is a 30% reduction in DDR memory bandwidth. This should give you an appreciation of why many video designs will chroma subsample and upsample many times — so that they can minimize precious memory and bandwidth resources.

Since in a majority of designs you will be dealing with 4:2:2 video we will use the second number of 7.44 Gbit/s.

Frame buffers are easy to calculate as they just read a frame and write a frame — the memory bandwidth calculation is simply 1920 × 1080 × 20 bits × 60 fps.

Now let's introduce a real-life constraint. We are using 20-bit YCbCr video data that has been sampled in the 4:2:2 sampling scheme, but what if we are using an external memory with 256-bit data bus? There is a mismatch.

This means that during each burst to the external memory, we are only able to transmit or receive 12 pixels (or 240 bits) total. This indicates that, for each transfer, 16 bits are wasted. Or you could describe this as 1.33 bits wasted for every pixel transmitted/ received.

In the same fashion, with 10-bit motion values, only 25 motion values can be transmitted/received during each burst. Therefore, for each transfer, 6 bits are wasted, or 0.24 of a bit is wasted for every motion value transferred/received.

Therefore we must recalculate the memory bandwidth for both the deinterlacer and the framebuffer as shown in Table 20.1.

After we figure out the penalties for each burst of video data and motion values, we are able to compute the worst-case external-memory bandwidth requirement for each IP. To do that, we first calculate the memory requirement for the worst-case input/output format and motion values. Then we sum up the bandwidth requirement for each write/read port. In this case, for each motion-adaptive deinterlacer, with the motion-bleed option on, the worst-case requirement is 7.909 Gbits/second. Each frame buffer also requires external memory bandwidth of 5.308 Gbits/second. It does not significantly change the numbers — but it is an important consideration when designing real-life video signal chains.

Since we have two video paths in the design that we started with, we need two deinterlacers and two frame buffers. The total worst-case system memory bandwidth requirement is 26.434 Gbits/sec, as shown in Table 20.2.

If you are using a DDR2 memory running at 267 Mhz, your theoretical peak memory bandwidth possible is:

266.7 MHz × 64 bits × 2 (both clock edges used) = 34.133 Gbit/s

# Table 20.1
## Calculating the memory bandwidth for the deinterlacer and the framebuffer

| Function | Worst Case Input Format | Worst Case Output Format | Worst Case Motion Format |
|---|---|---|---|
| Deinterlacer | (1080i60)<br>1920 × 1080 × 21.3 bits ×<br>60 / 2 fps<br>1.327 GBit/s | (1080p60)<br>1920 × 1080 × 21.3 bits ×<br>60 fps<br>2.654 GBit/s | (1 motion value per sample)<br>1920 × 1080 × 10.24 bits ×<br>60 / 2 fps<br>0.637 Gbit/s |
| Frame buffer | (1080p60)<br>1920 × 1080 × 21.3 bits ×<br>60 fps<br>2.654 Gbit/s | (1080p60)<br>1920 × 1080 × 21.3 bits ×<br>60 fps<br>2.654 GBit/s | N/A |

| Function | Memory Access | Bandwidth (GBit/s) |
|---|---|---|
| Deinterlacer | One read at input rate | 1 × 1.327 = 1.327 |
| | One read at motion rate | 1 × 0.637 = 0.637 |
| | One write at motion rate | 1 × 0.637 = 0.637 |
| | Two accesses at output rate | 2 × 2.654 = 5.308 |
| Total bandwidth | | 7.909 |
| Frame buffer | One write at input rate | 1 × 2.654 = 2.654 |
| | One read at output rate | 1 × 2.654 = 2.654 |
| Total bandwidth | | 5.308 |

# Table 20.2
## The total worst-case system memory bandwidth requirement

| Function | Bandwidth (Gbit/s) |
|---|---|
| 2 × deinterlacer | 2 × 7.909 = 15.818 |
| 2 × frame buffer | 2 × 5.308 = 10.616 |
| **Total** | **26.434** |

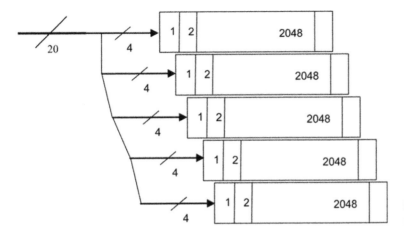

**Figure 20.6.** Five M9K RAM memory blocks in parallel.

Which means for this design you need an efficiency of 26.434/ 34.133 = 77.5%. Efficiency in this context is the effect of multiple masters trying to pull from the memory or write to it. And in some cases the deinterlacer memory access may be stalled if it is being used by the frame buffer. This would cause the entire processing to stall, so the memory subsystem has to be designed such that it meets the required efficiency.

## 20.3 Calculating On-Chip Memory

Video scaling will use the most on-chip memory. This calculation is fairly simple, but depends on the complexity of the scaling function.

For example, an important part of upsampling and downsampling is choosing the appropriate filter kernel. This can help preserve the sharpness of the edges during interpolation and avoid aliasing during decimation. Filter response aside, resource usage is another relevant and often-overlooked aspect of the decision making process. It is important to realize that a Nv × Nh filter kernel would translate into Nv + Nh multipliers and Nv line buffers.

Which means that if you choose a 4 × 4 filter kernel, you will need enough on-chip memory to store four lines of video. One video line-buffer stores all the pixels in a single line onto the FPGA memory. The size and the configuration of this video line buffer depends upon many factors.

As we have seen before, each pixel is generally represented by three color planes − RGB, YCrCb, etc. Typically each color plane in turn is encoded using 8, 10 or even 12 bits. We also have to factor chroma sub-sampling into our calculations.

# Table 20.3
## The number of bits required to store one line of video

| Video frame size (pixels) | # of pixels ina line | Color plane/bits per plane | Sub-sampling | Video line buffer (bits) |
|---|---|---|---|---|
| 640 × 480 | 640 | RGB/8 | 4:4:4 | 21 × 640 |
| 1280 × 72 | 1280 | RGB/8 | 4:4:4 | 21 × 1280 |
| 1280 × 72 | 1280 | YcrCb/8 | 4:2:2 | 16 × 1280 |
| 1920 × 10 | 1920 | YcrCb/10 | 4:2:2 | 20 × 1920 |

The number of bits required to store one line of video depends on multiple color space variables as shown in Table 20.3.

To minimize the number of memory blocks, it is extremely important to have the right memory configuration. Since high-definition is appearing in all facets of the video market, a line buffer size of 1920 pixels (typical HD resolution being 1920 × 1080) must be considered. Each pixel is generally chroma subsampled at 4:2:2, providing 20 bits per pixel.

The ideal configuration of a memory block when implementing a 1080p HD video line would therefore be 20 bits wide and 1920 bits (about 2K) deep.

Altera FPGAs have M9K RAM memory blocks that are designed to accommodate HD video. Each RAM memory block can be configured as 2K × 4 bits. Figure 20.6 shows that cascading five of these blocks in parallel enables a video line-buffer with a memory-bit efficiency of 93.75% (1920 / 2048) to be easily implemented.

When selecting FPGAs for HD video processing, the available configuration options of the embedded memory blocks will determine the number of video line buffers that can be implemented. A well-planned and flexible block RAM configuration leads to high bit-efficiency and will allow the design to fit into a smaller and more economical device.

## 20.4 Conclusion

When implementing HD video processing designs, memory resources are generally the more important consideration, given how memory-intensive video designs are. This chapter gives you a sense of what to expect both in terms of internal on-chip memory and also external DDR memory.

# 21

# DEBUGGING FPGA-BASED VIDEO SYSTEMS

**Andrew Draper, Altera video engineering**

## CHAPTER OUTLINE

**21.1 Timing Analysis   201**
 21.1.1 Check that the Design Meets Timing   202
 21.1.2 Fix Your Design if it Does Not Meet Timing   203
**21.2 The SystemConsole Debugger   204**
**21.3 Check That Clocks and Resets are Working   205**
**21.4 Clocked and Flow Controlled Video Streams   206**
**21.5 Debugging Tools   206**
**21.6 Converting from Clocked to Flow-controlled Video Streams   208**
**21.7 Converting from Flow-controlled to Clocked Video Streams   209**
**21.8 Free-running Streaming Video Interfaces   210**
**21.9 Insufficient Memory Bandwidth   211**
**21.10 Check Data Within Stream   212**
**21.11 Summary   213**

In this chapter we will discuss some of the strategies you can use for debugging a video system built in an FPGA. The examples use Altera's video debugging tools and methodology, although the concepts can be applied generally.

Before moving on to the video-specific parts of debugging it is worth checking that the design has synthesized correctly and has passed a number of basic sanity checks.

## 21.1 Timing Analysis

Hardware designs that run from a clock need to meet a number of timing constraints. The two most basic of these exist to prevent errors if a signal changes while it is being sampled by a register:

The input to a register must be stable for a time before the clock edge on which it is sampled — referred to as the *setup time*.

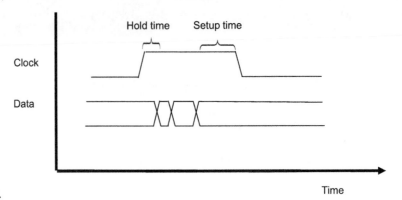

**Figure 21.1.** Setup and hold times.

The input to a register must remain stable for a time after the clock edge on which it is sampled — referred to as the *hold time*.

Most signals originate from registers in the same clock domain, the outputs of which change just after the clock edge (i.e. there is a delay going through the register). There are also delays while the signals pass through combinational logic, and further delays if the signals need to be routed across the chip to their destination. The sum of these delays is known as the *propagation delay*.

The mathematical relationship between the delays is expressed by the following two equations which must be satisfied for all paths within the chip:

propagation delay + setup time < = clock period
propagation delay > = hold time

Where < = is the mathematical less than or equal symbol, and > = is greater than or equal

There are more complex timing issues when signals cross from one clock domain to another but these are usually handled by specially designed library components.

A hardware design where these equations are satisfied for all signals on the chip is said to *meet timing*. A design which does not meet timing will usually fail in subtle and unexpected ways so further debugging is not usually productive.

### 21.1.1 Check that the Design Meets Timing

During synthesis the layout tool will place the logic within the chip and then run a timing analysis to check that the design meets the setup and hold requirements of the chip that will implement it. If these requirements are not met, the tool will adjust the layout and run the timing analysis again, continuing until timing analysis passes.

For the timing analysis stage the designer must provide scripts to tell the tool what timing behavior is required. These scripts are written by the hardware designer and shipped with the library component (if you write your own hardware with multiple clock domains then you will need to provide these scripts). If these scripts are incorrect, or if the clock speeds set in the scripts are lower than the actual speed of the clock, then the design can fail even if it meets timing.

A timing failure in one part of the circuit can cause problems elsewhere in the design, because if one part of the design fails to meet timing then the tool will stop rearranging the design throughout the chip. It will report the errors that have caused it to stop processing, but may suppress errors for other areas of the design which have not been completely processed. Thus a timing failure in one part of the chip is said to hide failures elsewhere in the design.

The propagation delay varies with several factors:

The temperature of the silicon within the chip: recent chips run fastest at room temperature and slowest at the top and bottom ends of their temperature range.

Manufacturing variations can change the propagation delay between one batch of chips and the next. Manufacturers partly deal with this by measuring the speed of chips after production and assigning a higher speed grade (and price) to those with lower propagation delays but there is still a small variation within each speed grade.

Small changes in the supply voltage: tolerances within the power supply components allow for difference between the supply voltages from one board to another.

The timing analysis tool will usually check the timing multiple times with different timing modes — for example it will check both the maximum and minimum propagation delays for the temperature and manufacturing variation.

All timing models for a design must pass before it can be used in a production system is used when timing passes only at lower temperatures: a liberal application of freezer spray to the chip can make a design work for a minute or two — often long enough to indicate that timing is the cause of failures.

## 21.1.2 Fix Your Design if it Does Not Meet Timing

Here we will be referring back to the two basic timing equations above. In most chips the propagation delay is significantly larger than the hold time, so the first equation is the harder to satisfy. This equation can always be satisfied by decreasing the clock frequency (increasing the clock period) but this is usually unsatisfactory, especially for video designs where a minimum clock frequency is required to process all the pixels in a frame.

Another method is to buy a chip at a faster speed grade. Unfortunately this has cost implications or is not possible because previously shipped products need to be upgraded.

Other methods of making a design meet timing include: inserting buffer registers to reduce the length of combinational paths; changing the layout to place critical registers closer to each other; and reducing the fan out of signals (which can increase switching speed and make the layout simpler).

If the timing tool reports hold-time violations that reducing the clock speed will not fix, design changes are required. Refer to FPGA Design: Best Practices for Team based design (Simpson) ISBN−13: 978-1441963383.

If you are using library components to create your design then the component designer will have already considered these issues and may have included parameters which help their component meet timing (usually in exchange for an increase in size or latency).

Many libraries include components called *pipeline bridges* (or similar names) which can be used to easily insert buffer registers into all the signals of a bus without affecting its behavior.

## 21.2 The SystemConsole Debugger

As we will use SystemConsole as an example of a tool running on a debug host we will now provide a basic introduction.

Debug tools usually refer to the system being debugged as the *target* − the system which you use to debug the target is the *debug host*. The host will be connected to the target via one or more debug cables (nowadays these are normally JTAG, USB or Ethernet − though debugging over other media is possible).

To enable debugging, the system designer places *debug agents* within the target system. These agents are sometimes packaged within other components − for example, most processor components now contain a debug module − or they can be explicitly instantiated by the system designer.

Those debug agents that use a JTAG interface to communicate with the host are automatically connected to the JTAG pins on the device by the Quartus software. In the current Altera software, debug agents using other cables (USB and Ethernet) must be explicitly connected to the pins on the device.

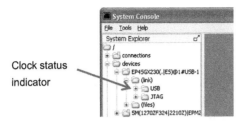

Clock status
indicator

**Figure 21.2.** Clock sense indication.

# 21.3 Check That Clocks and Resets are Working

Incorrectly functioning clocks or resets are a common cause of design failures, which should be ruled out early in the debug process — even experienced engineers have wasted hours of time debugging apparently failed systems where the clock has been disabled or the wire supplying the clock signal from a test device to the board has been knocked off.

Other causes of clock failures include Phase locked loops which are unable to lock because their input signal has too much jitter or is outside the acceptable range of input frequencies.

Reset signals can also become stuck — either holding part of the design in reset permanently or never resetting it. If a design is not reset then it does not start in a consistent state, and may get into a state that its designer did not intend. Sometimes a design will get out of these unusual states and sometimes it will become stuck.

FPGA designs with reset faults sometimes work because the configuration logic within the FPGA sets most registers to their defined reset state at the end of configuration.

Most debug tools, for example the Altera SystemConsole tool, provide ways to check that clocks are running and resets are behaving correctly. In SystemConsole the explorer window shows a green clock badge on nodes that have a running clock and a red clock badge (with associated tooltip) on nodes which can sense the clock but do not detect it running.

It also provides the jtag_debug service to give scripted access to the clock sensing hardware. The TCL below shows an example of its use:

```
set jd [lindex [get_service_paths jtag_debug] 0]
open_service jtag_debug $jd
puts "Clock running: [jtag_debug_sense_clock $jd]"
puts "Reset status: [jtag_debug_sample_reset $jd]"
```

## 21.4 Clocked and Flow Controlled Video Streams

As you have read in earlier chapters most digital video protocols send video frames between boards using a clock and a series of synchronization signals. This is simple to explain but it is an inefficient way to communicate within a device, as all processing modules need to be ready to process data on every clock within the frame, but will be idle during the synchronization intervals.

Using a flow-controlled interface is more flexible because it simplifies processing blocks and allows them to spread the data processing over the whole frame time. Flow-controlled interfaces provide a way to control the flow of data in both directions – the source can indicate on which cycles there is data present and can backpressure when it is not ready to accept data. In the Avalon ST flow-controlled interface the valid signal indicates that the source has data and the ready signal indicates that the sink is able to accept it (i.e. is not backpressuring the source).

If you are building a system from library components, most problems will occur when converting from clocked-video streams to flow-controlled video streams, and vice versa.

## 21.5 Debugging Tools

Several debugging tools are available: the most basic tools of which are an oscilloscope, logic analyser or (within an FPGA) an embedded logic analyser (such as Altera's SignalTap tool). These tools provide a high-resolution view of the data being transferred on a number of signals.

If you have data integrity issues between boards, then low-level debugging tools such as these can be used to diagnose the problem. Unfortunately, once the signals between boards or within devices are clean, these tools typically provide too much data to diagnose the types of problems that appear at higher levels.

Higher-level debug tools provide a way to trace the data passing through the system and display it as video packets. The amount of data in a video system is more than can easily be transferred, so it must be compressed to allow it to be transferred to the debug host and analysed.

The highest level of compression can be achieved by ignoring most of the pixel values and only transferring control packets and statistics about the data flow – for example, a count of the number of clock cycles where data was transferred, was not available to transfer from the source or was back-pressured by the sink.

The Altera trace system is instantiated when you are building a video design within the QSYS environment. Two parts are needed: a trace monitor component for each interface to be traced and a trace system component which transfers trace data packets to the host.

A video trace monitor component needs to be inserted into each video stream you want to monitor. This component is non-intrusive – it has no effect on the video data going through the stream. The video trace monitor component reads the signals being transmitted and sends summaries to the trace system. You will need to parameterize the video trace monitor to match the type of data being sent and to match the trace system data-width.

The trace system component takes the reports from the trace monitors and buffers them before sending the results to the host

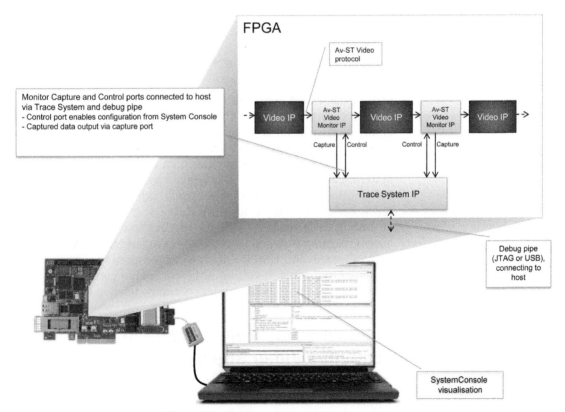

**Figure 21.3.** Trace monitors and the trace system.

**Figure 21.4.** Example of decoded trace output.

over JTAG or USB, where they are reconstructed for the user. You will need to parameterize this component to select the type of connection to the host, the number of monitors, the buffer size, etc.

The SystemConsole host application decodes and displays the received packets to show the data as it passes through the system. Each video packet is displayed as one line in the display. The sections below describe common video errors and how to recognize them in the trace output.

Debug tools are also available which allow the debug host to access memory mapped slaves within the target. The Altera JTAG Avalon Master and USB Debug Master components are explicitly designed to do this: if you do not have such a component available then most processor debuggers can be used in a similar way.

## 21.6 Converting from Clocked to Flow-controlled Video Streams

In a functioning system the input to the flow-controlled domain will send data as it becomes available. The system needs to transfer, on average, one line's worth of pixels in each line scan time. The transfer of data will normally be controlled by "valid", with "ready" asserted occasionally to select the cycles on which data is accepted.

The number of cycles on which "valid" is asserted depends on the ratio between the screen resolution and the clock rate in the flow-controlled domain. If the clock is just sufficient for the highest resolution then "valid" will be asserted on most cycles within the main part of the frame. At lower resolutions "valid" will only be asserted on a proportion of the cycles.

The "ready" signal to the clocked video input should not be the main source of flow control on the frame, so it is typically de-asserted only for short periods to synchronise with the sink. One common problem is that if "ready" is de-asserted for too long then the memory buffer in the video input block can overflow.

Attaching a streaming video monitor to the output of the video input block can help detect overflow situations — if the video input block is backpressured (by de-asserting "ready") for too long then it will abandon the backpressured frame and send a short packet. This can be seen on the trace.

The trace also reports the number of not-ready cycles within each packet and the time interval between packets. This can be used to check that the interface is being mostly flow-controlled by "valid" rather than "ready".

If the clocked video-input block has a control port then the debug master can be used to check the overflow sticky-bit in the status register. This bit will be set if there has been an overflow since it was last checked - note that if you have software monitoring and clearing this bit then reading it from the debugger will not be reliable.

## 21.7 Converting from Flow-controlled to Clocked Video Streams

The clocked-video output component converts flow-controlled video packets into a clocked video signal. The flow control on the input to this component is controlled by the "ready" signal, which essentially pulls data out of the interface as it is needed.

If the source is unable to provide data at a sufficient rate then the FIFO in this component will empty. This is referred to as *underflow*. At this point the component tries to re-synchronize, sending out blank video data and reading continuously from the input until the start of the new frame appears, when it will re-start the output video.

The clocked-video output component latches the underflow indication — the underflow sticky-bit is set when an underflow occurs. You can use the debug master or software on an embedded processor to check this bit. As with the overflow bit in the clocked video master, if embedded software is monitoring and resetting the bit then reading it from the debugger will not be reliable.

The video trace monitor can also indicate when there are problems with underflow. Normally the stream going in to the clocked-video output is controlled by "ready", but if there is a problem with underflow then "ready" will not be asserted during the re-synchronization process. The resulting lack of backpressure is visible in the captured video packet summaries.

# 21.8 Free-running Streaming Video Interfaces

The clock rate within a flow-controlled video system is normally set to sufficient bandwidth on the streaming ports for a picture of maximum resolution to be transmitted (with a small amount of overhead to allow for jitter).

The flow control signals — "ready" near the video output or "valid" near the video input — ensure that processing does not run faster than the incoming video stream. If processing runs too far ahead then frames will be missed and the picture will be jerky.

This can happen if the design has instantiated multiple triple-buffer components. Triple buffers do not flow-control their inputs or their outputs (except temporarily when waiting for memory accesses). A video pipeline between two triple buffers will run at the processing clock speed rather than staying in sync with the video frames.

If part of the video pipeline is allowed to free run then this will waste memory bandwidth. It can also reduce picture quality as the input triple buffer will duplicate frames to keep its output busy while the output triple buffer will delete frames to match the frame rate on the output. The overall effect will be that some frames are output multiple times while other frames are not output at all.

The solution to the free-running problem is to replace all but one of the triple buffers with a double-buffer component. The double buffer does no frame rate conversion so will not allow its input and output to run more than one frame apart. This will provide flow control to the central part of the system.

The video trace monitor can also be used to detect free-running streaming video components. Examining the flow-control statistics will report that there is no backpressure or unavailable data – i.e. "ready" and "valid" will be high for most of the frame.

The timing information on the captured video packets reports the average frame rate passing through the monitor. If the streaming video interface is free running then the frame rate in parts of the video pipeline will be much faster than expected.

# 21.9 Insufficient Memory Bandwidth

Some video processing components, such as a color space converter, can process the video data one pixel at a time. Others need to store the pixels between input and output – the simplest examples are the buffer components that write the input pixels to a frame buffer in memory and read from memory (with different timing) to create the output pixel stream.

Components using a frame buffer demand a large amount of memory bandwidth – the sum of the bandwidth of the input and output data rates. If the memory subsystem is not designed correctly then it will not be able to provide this bandwidth. This will cause excessive flow control of the input and/or output which in turn will make FIFOs in other components overflow or underflow as described previously.

Because of their size, most frame buffers are stored in external memory, which is usually shared between multiple, different, memory-mapped masters. Even in the case where memory is not shared, a double- or triple-buffer component has two masters, one to write and the other to read.

When there are multiple masters for the same memory-mapped slave, an arbiter is needed to share the slave's bandwidth between the masters. In some cases the arbiter is inserted automatically as part of the bus fabric – in other cases it is explicitly inserted by the user as a separate component, or as part of the slave component.

The Altera multi-port front-end component is a specialized arbiter which understands the costs of different DDR accesses and can be configured to maximize bus efficiency. When used correctly this component can achieve memory-bandwidth efficiency of over 90% – i.e. the number of cycles lost due to bank opens, closes, read-after-write delays and other DDR performance hazards is less than 10%.

Setting up the arbiter to achieve high efficiencies is sometimes complex, as the interface priorities need to be set correctly so that low-latency masters are serviced quickly. Most video component masters will use only as much bandwidth as is needed for the selected video resolution, although in example of free running they will use as much bandwidth as available — possibly locking out lower-priority masters from the memory.

A processor master does not normally have a bandwidth limit — it will use as much bandwidth as is available and can be consumed. Most processors are only able to pipeline a limited number of memory accesses, so the latency of the memory can limit the amount of bandwidth they can consume. Processors are normally put at the lowest priority to prevent them from starving video masters, which have a bandwidth target.

Many arbiters, including the Altera external memory interface toolkit, have an optional efficiency monitoring feature which collects statistics about the bandwidths and latencies used by different masters. This efficiency monitor can be used to check that the memory is running at a sufficiently high overall bandwidth, and can help with optimization when it is not.

## 21.10 Check Data Within Stream

During the prototype stage all components have bugs that must be fixed. The usual hardware flow is to fix these bugs through simulation where the visibility into the system is good.

This is harder for video components as the high data rates mean that complex components can take several minutes to simulate each frame. For edge-case bugs, which occur once every few hours on video data, this would mean many days of simulation before a bug occurs. These bugs are only really debuggable in hardware.

Most debug components, including the Altera trace system, can be set up to continuously capture data into a circular buffer. When the trace system is *triggered* it stops capturing data, sending its stored data to the host for analysis. Ignoring the activity of the system significantly before the trigger lets you concentrate on the immediate causes of the bug, rather than having to wade through large amounts of captured data.

What drives the trigger signal? For hard-to-find bugs you might write custom hardware, which monitors various parts of the system and sends a trigger when misbehavior is detected. This is difficult, can be error-prone and is not always necessary.

Most component vendors ship bus protocol monitors that are used in simulation to check that the signals on a bus do not violate the specification. For example, a lot of memory-mapped buses require that after an access has started the address signals must remain stable until that access is accepted by the slave. A master that changes the address lines halfway through its transaction will be detected by the monitor.

Bus monitors are used extensively during simulation and some of them are now synthesizable, so they can be temporarily included within an FPGA design. Connecting the error output of the monitor to the trigger input of a trace system, or embedded logic analyzer, will let the user capture the events leading up to, and just after, the error.

In streaming video systems application-aware bus monitors also detect higher-level errors: for example, a component which outputs data packets which do not match the size described in the preceding control packet will be logged and/or report errors.

The video trace system will show edge cases occurring which might trigger a bug, for example when two control packets preceding a data packet is legal (the second control packet takes priority) but is not handled correctly by some components.

## 21.11 Summary

Debugging video systems can be daunting, especially when the only visible symptom is the output of a black picture.

Many trace components are available to provide visibility into the system and narrow down the location of the bug which is causing the symptom. Careful use of these components can save significant time during development.

In some cases the trace components can be left active in shipped systems. Remote debugging can then be used on units running real data – this can be especially valuable when the bug has been triggered by almost standard data being generated by other equipment that is only installed in one broadcaster, in one distant country.

# INDEX

*Note*: Page numbers followed by "f" and "t" indicate figures and tables respectively

## A

ADC. *See* Analog-to-digital converter
Advanced Video Codec (AVC), 126
Aliasing, 14
  *See also* Nyquist sampling rule
  spinning wheel experiment. *See* Spinning wheel experiment
Alpha blending, 49
  alpha frame, 50
  alpha values, 50
    background pixel, 52
    composite frame, 52
    physical background, 51–52
    subject sits, 52
    video processing, 52
  blend frame, 50
  composite frame, 50
  composite pixel, 50
  hardware implementation, 51
    background pixel, 51
    composite pixel, 51, 51f
    composite pixel color components, 51, 52f
    composite pixel creation, 51
    using equation, 51
    foreground pixel, 51
    multiplier in hardware, 51
    pixel value, 51
    re-arrangement, 51
  logo from frames, 50
  math behind, 50
  math value calculations, 50
temporal processing, 49
translucent, 50
zipping past, 49–50
Altera trace system, 207
Altera's DSPbuilder toolflow, 165f
Analog signal, 1
Analog-to-digital converter (ADC), 11
Anamorphic widescreen, 38
Application specific standard product (ASSP), 29
Arbitrary slice ordering (ASO), 139
ASO. *See* Arbitrary slice ordering
ASSP. *See* Application specific standard product
Asynchronous transfer mode (ATM), 174
ATM. *See* Asynchronous transfer mode
AVC. *See* Advanced Video Codec

## B

B-frames, 115–116
Baseline Profile (BP), 131–133
Bayer demosaicing, HW implementation, 55
  Bayer data, 56
  bilinear implementation, 56
  FPGA, 56
  FPGA structure, 56f
Bayer filters, 12
Bayer to RGB conversion. *See* Bayer demosaicing
Bicubic scaling, 32, 33f

Bilinear interpolation technique, 56
Bilinear scaling, 31, 31f
Bit narrower, 33–34
Bits, 7t
  in flat-panel TV, 7
  in HD video, 7
Blend frame, 50
Block artifacts, 144
  applying color planes, 145
  deblocking filter, 145f
  filter, 144
  H.264 deblocking filter, 145
  quantization level, 144
  video content, 144
Block size. *See* Macroblock
Bob deinterlacing, 41, 43–44
  with scan-line duplication algorithm, 45
  with scan-line interpolation algorithm, 45
BP. *See* Baseline Profile
Bursty video stream, 193

## C

CABAC. *See* Context-based adaptive binary coding
Cadence, 47
Cadence detection standard, 47
  cadence detection, 47–48
  deinterlacers, 48
  frames and fields, 47
  motion picture photography, 47
  Panasonic, 47

Cadence detection standard
(*Continued*)
  3:2 pull-down technique,
    47
  3:2 video cadence, 47
  24 fps film, 47
Cathode ray tube (CRT), 6
CAVLC. *See* Content-based
    adaptive variable-length
    coding
CBP. *See* Constrained Baseline
    Profile
CCD. *See* Charge coupled
    device
CFA. *See* Color filter array
Charge coupled device (CCD), 11
  sensors, 53
CMOS sensors. *See*
    Complementary metal
    oxide semiconductor
    sensors
CMOS technology. *See*
    Complementary
    metal oxide
    semiconductor
    technology
Color filter array (CFA), 54
Combing effect, 41
Communication systems,
    162–163
Complementary metal oxide
    semiconductor sensors
    (CMOS sensors), 53–56
  Bayer demosaicing, 55
  Bayer image, 54
  bilinear interpolation
    technique, 56
  color filter array, 54, 54f
  color plane values, 55, 55f
  8MP, 53
  missing color information
    calculation, 55
  missing color plane
    calculation, 55
  optimum quality image, 53
  pixel filter array, 54, 54f
  pixel sensors, 54
  processing pipeline functions,
    53

sensor and video processing,
    54f
  standard video format, 53
  video scaling, 55–56
Complementary metal oxide
    semiconductor
    technology (CMOS
    technology), 11
Component video, 65
  cables, 65f
Composite frame, 50
Composite pixel, 50
Composite tiling, 109
Composite Video Blanking and
    Sync (CVBS), 64
  composite cable, 64f
Compression artifacts. *See*
    Block artifacts
Constellation, 147–148
  QPSK, 148f
  16-QAM, 148, 149f
  16-QAM recovered, 161f
  64-QAM recovered, 162f
  size and bit rate, 148t, 149t
  trajectory, 151f
Constrained Baseline Profile
    (CBP), 131
Content-based adaptive
    variable-length coding
    (CAVLC), 138
Context-based adaptive binary
    coding (CABAC),
    138–139
Contour integral engine,
    190f
Conventional digital filter,
    141
CRT. *See* Cathode ray tube
CVBS. *See* Composite Video
    Blanking and Sync

**D**

DAC, 166, 169
dB. *See* Decibels
DCT. *See* Discrete Cosine
    Transform
Deblocking filter, 145
Debug agents, 204
Debug host, 204

Debugging FPGA-based video
    systems
  clock sense indication, 205,
    205f
  clocked to flow-controlled
    video stream conversion
    data transferring, 208–209
    debug master, 209
    ready signal, 209
    valid signal, 209
  data check within stream, 212
    for edge-case bugs, 212
    trigger signal, 212
    video trace system, 213
  debugging tool. *See*
    Debugging tools
  flow-controlled interface, 206
  flow-controlled to clocked
    video stream conversion
    ready signal, 209
    underflow, 209–210
  free-running streaming video
    interfaces, 210–211
  memory bandwidth, 211
    Altera multi-port front-end
      component, 211
    arbiters, 211–212
    external memory, 211
  resets signals, 205
  System Console debugger, 204
  timing analysis. *See* Timing
    analysis
Debugging tools, 206
  Altera trace system, 207
  higher-level, 206
  System Console host
    application, 208
  video trace monitor
    component, 207, 207f
Decibels (dB), 79
  logarithmic expression, 80
  negative, 81
  Ohm's law equation, 80
  power measurements, 80
    amplitude ratio doubling, 81
    power ratio doubling, 81
Decode timestamp, 171–172
Deinterlacer, 46, 48, 194–195,
    195f, 196f

Deinterlacing techniques, 40
  bob deinterlacing, 41
  combing effect, 41, 42f
  fields and frames, 40, 41f
  image changes, 41
  image intensity and vertical
      resolution, 43
  interpolation, 41
  motion-adaptive
      deinterlacing, 43
  scan-line duplication, 42,
      43f
  scan-line interpolation, 42,
      43f
  spatial line doubling, 41
  weave deinterlacing, 40–41,
      42f
DFT. *See* Discrete Fourier
      Transform
DHCP. *See* Dynamic host
      reconfiguration
      protocol
Differential encoding, 74–75
Digital cinema, 140
Digital filtering, 19, 159
  FIR filtering. *See* Finite
      Impulse Response filters
  median filtering. *See* Median
      filter
Digital subscriber lines (DSL),
      174–175
Digital video
  bits, 6–8
  color space, 8
  pixels, 5–6
  resolutions, 5
  RGB to YCrCb conversion, 9
Digital Video Interface (DVI),
      63
  connector, 63f
Discrete Cosine Transform
      (DCT), 83, 143
  basis functions, 100, 100f
  coefficients, 98
  cosine frequencies in, 97–98,
      99f
  equation, 98
  pixels uses, 97–98, 100f
  purpose, 100–101

Discrete Fourier Transform
      (DFT), 83
  *See also* Fast Fourier
      Transform (FFT)
  equations, 85–86, 86t
  first DFT example, 86–87
    using IDFT, 87–88
    non-zero term, 87
  fourth DFT example, 90–93
  second DFT example,
      88–89
  third DFT example, 89–90
Display port, 61–62, 62f
Double buffering, 192
Double frame buffer, 192
Down conversion, 152
Download and play, 176
DSL. *See* Digital subscriber
      lines
DSPBuilder, 168
DVI. *See* Digital Video Interface
Dynamic host reconfiguration
      protocol (DHCP), 170

**E**

8MP CMOS sensor, 53
Electro-optical infrared system
      (EOIR system), 57
Elementary stream, 171
Encapsulation, 171
End of block symbol (EOB
      symbol), 106
Entropy, 69–70
EOB symbol. *See* End of block
      symbol
EOIR system. *See* Electro-
      optical infrared system
Ethernet, 174
Euler equation, 84
Extended Profile (XP), 133

**F**

Fast Fourier Transform (FFT),
      83
  effectiveness, 96
  flow graph, 96f
  generic DFT equation, 93
  radix 2, 97

FFT. *See* Fast Fourier
      Transform
Field programmable gate array
      (FPGA), 9
  implementation, 190
Fields per second (fps), 41
Finite Impulse Response filters
      (FIR filters), 19–20
  construction, 20
    convolution, 21, 23
    frequency response, 21
    small 5-tap parallel filter,
        20, 21f
  frequency response
      computation, 23–24
    complex exponential input,
        25
    complex exponential signal,
        25
    5 tap filter, 26
    gain of filter, 26
    logarithmic difference,
        28
    magnitude, 27
  structure, 21
FIR filters. *See* Finite Impulse
      Response filters
Focal plane array (FPA), 57
Focus-score assessment,
      181
Fourier transform, 84–85
FPA. *See* Focal plane array
FPGA. *See* Field programmable
      gate array
fps. *See* Fields per second;
      Frames per second
Frame buffer, 191–192
  bursty video stream, 193
  deinterlacers, 194
  double, 192, 192f
  down-scaler output, 194,
      194f
  frame dropping and
      repeating, 193
  memory transactions,
      194–195
  triple, 192–193, 192f
  2-D median filters, 194
  video frame, 193, 193f

Frame buffer (*Continued*)
  writer and reader block,
    192
Frame processing order,
    114–116
Frames per second (fps), 1

## G

GOP. *See* Group of pictures
Gradient technique, 182–184
Group of pictures (GOP), 115,
    116f

## H

H.264 deblocking filter, 145
H.264 video compression
    standard, 131
  ASO, 139
  CABAC, 138–139
  CAVLC, 138
  deblocking filter, 137
  features, 139–140
  flexible macroblock
    ordering, 139
  frequency dependent
    quantization, 135–136,
    136f
  levels, 131, 132t
  logarithmic quantization, 135
  MBAFF, 138
  pixel interpolation, 136
  profiles, 131, 134t
    BP, 131–133
    CAVLC 4:4:4 intra profile,
      134
    CBP, 131
    Hi10P, 133
    Hi422P, 133
    Hi444PP, 133
    high 10 intra profile, 134
    high 4:2:2 intra profile,
      134
    high 4:4:4 intra profile, 134
    HiP, 133
    MP, 133
    PHiP, 133
    XP, 133
  redundant slices, 139

spatial prediction, 138
support for 4 x 4 integer DCT,
    135
temporal prediction, 137–138
variable block size motion
    estimation, 136–137, 137f
HD. *See* High definition
HDMI. *See* High-definition
    multimedia interface
HDPC. *See* High definition
    digital content
    protection
High 10 Profile (Hi10P), 133
High 4:2:2 Profile (Hi422P), 133
High 4:4:4 Predictive Profile
    (Hi444PP), 133
High definition (HD), 5
  video processing, 195
High definition digital content
    protection (HDPC),
    61–62
High Profile (HiP), 133
High-definition multimedia
    interface (HDMI), 63
  connector, 63f
HiP. *See* High Profile
Hold time, 202, 202f
Horizontal synchronization
    signal, 1
HTTP. *See* Hypertext transfer
    protocol
Huffman coding, 71, 108
Hypertext transfer protocol
    (HTTP), 176

## I

I-frames, 115
IDFT. *See* Inverse Discrete
    Fourier Transform
IGMP. *See* Internet group
    management protocol
Image compression
  baseline JPEG, 103–104
  DC scaling, 104
  entropy encoding scheme,
    106–107
  extensions, 108–109
  quantization table, 104–106

Image scaling, 57
Image sensors, sensor
    processing for
  *See also* Joint Photographic
    Experts Group (JPEG);
    Military EOIR system,
    sensor processing in
  Bayer demosaicing, HW
    implementation. *See*
    Bayer demosaicing, HW
    implementation
  CCD sensors, 53
  CMOS sensors. *See*
    Complementary metal
    oxide semiconductor
    sensors
  high-performance sensor
    processing, 59
  image sensor data functions
    digital binning, 58
    digital zoom, 58
    electronic image
      stabilization, 59
    local-area adaptive contrast
      enhancement, 58
    motion detection, 59
    multi-target tracking, 59
    noise filtering, 58
    non-uniformity correction,
      58
    pixel-level adaptive image
      fusion, 59
    super-resolution, 59
    wide dynamic range
      processing, 58
Image tiling, 108
  composite tiling, 109
  pyramidal tiling, 109
  simple tiling, 108
Image-processing system,
    181
  focus measurement, 182
    focused and unfocused
      images, 182f
    gradient techniques,
      182–184
    variance techniques,
      184–185
    vertical edge, 183f

focus score, 181
   assessment, 181
   segmentation algorithm. *See*
      Segmentation algorithm
Imagette image, 109
Input decoder buffer, 121,
   121f
Intellectual Property (IP), 34
Inter-symbol interference
   (ISI), 154
Internet, 175
Internet group management
   protocol (IGMP), 178
Internet protocol (IP), 169
   addresses, 170
   header, 169
   networks, 172
   packet formatting, 170f
   routers, 170
Internet protocol transport (IP
   transport), 173
   aggregated access to, 175f
   ATM, 174
   central office, 174–175
   DSL, 174–175
   Ethernet, 174
   LAN, 174
   SONET, 174
   technology, 175
   WAN, 174
Interpolation, 41, 68
Inverse Discrete Fourier
   Transform (IDFT),
   83–84
   equations, 85–86
IP. *See* Intellectual Property;
   Internet protocol
IP transport. *See* Internet
   protocol transport
ISI. *See* Inter-symbol
   interference

**J**

Joint Photographic Experts
   Group (JPEG), 103
   *See also* Image-processing
      system; MPEG-2
   baseline, 103–104

DC scaling, 104
encode and decode steps, 107,
   107f
entropy encoding scheme,
   106, 106f
extensions, 108
   image tiling, 108
   selective refinement, 108
   variable quantization, 108
Huffman coding, 107
quantization table, 104
   lossy compression, 105–106
   output array, 105
   value, 105
Joint Video Team (JVT), 126
JPEG. *See* Joint Photographic
   Experts Group
JPEG Tiled Image Pyramid
   model (JTIP model), 109
JTIP model. *See* JPEG Tiled
   Image Pyramid model
JVT. *See* Joint Video Team

**L**

LAN. *See* Local area network
Last mile, 174–175
Line buffer, 33
Line-of-sight system, 150
Local area network (LAN), 174
Lossless compression, 75–76
Lossy compression, 105–106
Low-power FPGA, 58

**M**

MAC. *See* Media access
   control; Multiply-and-
   accumulate
Macroblock, 112, 113f
Macroblock-adaptive frame-
   field (MBAFF), 140
Macrocell, 115
MAD. *See* Minimum absolute
   differences
Main Profile (MP), 133
Markov source, 72–73
Math behind alpha blending,
   50
MBAFF. *See* Macroblock-
   adaptive frame-field

Media access control (MAC),
   174
Median filter, 19, 20f, 142, 142f
   smoothing technique, 19
Meet timing, 202
MEMS. *See* Micro-electro-
   mechanical-systems
Micro-electro-mechanical-
   systems (MEMS), 185
Military EOIR system, sensor
   processing in, 57
   FPA algorithm, 57
   FPGA-based camera
      cores, 57
   high-quality sensor control,
      57–58, 57f
   low-power FPGA, 58
   military imaging systems, 56
   NVG, 58
   pixel streams, 57
   WDR processing, 57
Minimum absolute differences
   (MAD), 112
Minimum mean square error
   (MMSE), 112
MMSE. *See* Minimum mean
   square error
Morse code, 71
Mosiac, 109
Mosquito noise, 143, 143f
   *See also* Salt-and-pepper
      noise
   computer generated graphics,
      144
   DCT, 143
   sampling theorem, 144
   spatial image analysis, 144
Motion bleed, 44–45
Motion estimation, 114
   B-frames, 115–116
   GOP, 115, 116f
   I-frames, 115
   macrocell, 115
   P-frames, 115
   pixel block, 115
   residual artifacts, 114–115
   slice, 115
Motion picture photography,
   47

Motion-adaptive deinterlacer, 195
Motion-adaptive deinterlacing, 43, 45
  bob deinterlacing, 44
  using formula, 44
  infinite motion, 44
  missing pixel value calculations, 43–44, 44f
  motion bleed, 44–45
  motion calculation, 44–45, 45f
  3 × 3 window fields, 43, 44f
  weave deinterlacing, 44
Mouse teeth effect, 41, 42f
Moving Pictures Experts Group (MPEG), 113
MP. *See* Main Profile
MPEG. *See* Moving Pictures Experts Group
MPEG-2, 125
  aspect ratio support, 130
  field DCT, 129, 130f
  frame-type DCT, 129, 130f
  interlaced video support, 129
  levels, 126–127, 127t
  motion estimation, 129
  motion prediction, 129
  permutations, 128, 128t
  profiles, 127, 127t
  quantization, 129
  support for
    3:2 pulldown, 130
    4:2:2, 128–129
Multi-tap filter, 55
Multicast video, 177–178
Multicasting, 177–178
Multiply-and-accumulate (MAC), 34

**N**

NAL. *See* Network Abstraction Layer
Narrow filter impulse response, 154f
NAT. *See* Network address translation

NCO. *See* Numerically controlled oscillator
Nearest neighbor method, 30–31
Network Abstraction Layer (NAL), 139–140
Network address translation (NAT), 170
Night-vision goggle (NVG), 58
Noise quantization, 77
Non-internet transmission, 175
Non-line-of-sight system, 150
Non-linear filtering, 141
Numerically controlled oscillator (NCO), 167
NVG. *See* Night-vision goggle
*Nyquist* frequency, 16–17
Nyquist sampling rule, 17f
  *See also* Aliasing
  microphone signal, 17
  *Nyquist* frequency, 16–17
  quantization, 18

**P**

P-frames, 115
Packet identifier (PID), 172
Packetized elementary streams (PES), 171
PAL video source, 46
Passband signal, 158–159
PES. *See* Packetized elementary streams
PHiP. *See* Progressive High Profile
Picture Parameter Set (PPS), 139–140
PID. *See* Packet identifier
Pipeline bridges, 204
Pixel, 5, 6f
  block, 115
  color, 6–7
  filter array, 54, 54f
Polyphasing, 164f, 167
PPS. *See* Picture Parameter Set
Predictive coding, 73–74, 111–112
Presentation timestamp, 171–172

Program stream, 171
Progressive download, 176
Progressive frame, 39
Progressive High Profile (PHiP), 133
Propagation delay, 202
Pulse shaping filter, 152, 158
  frequency spectrum, 154–155
  I values, 152
  inter-symbol interference, 154
  narrow filter impulse response, 153–154, 154f
  Q values, 152
  rate calculation, 152
  receiver symbol-detection performance, 155
  relationship, 152
  sampling, 155
  sinc impulse response, 152–153, 153f
Pyramidal tiling, 109

**Q**

QAM. *See* Quadrature amplitude modulation
QPSK. *See* Quadrature phase shift keying
Quadrature amplitude modulation (QAM), 148
Quadrature phase shift keying (QPSK), 147
Quantization, 76
  *See also* Entropy; Predictive coding
  actual error level, 76–77
  complexity, 79
  dynamic range, 77
  noise, 77
  SNR, 77–78
  uniform quantizing, 78
  vector, 79
Quantization scale factor, 120
  input decoder buffer, 121, 121f
  Mquant, 120–121
  VBV model, 121–122
  video compression. *See* Video compression

**R**

Raised cosine filter, 155
  *See also* Pulse shaping filter
  acquisition process, 161
  Altera's DSPbuilder toolflow,
    165f
  digital filtering, 159
  digital transmit circuit path,
    163f, 166
  DSPBuilder, 168
  frequency response, 156f
  frequency rolloff table, 155t
  ideal low-pass filter, 158
  modulation and
    demodulation, 162
  numerically controlled
    oscillator, 167
  passband signal, 158–159
  polyphasing, 164f
    using polyphasing
      techniques, 167
  pulse shaping filter, 158
  raised cosine impulse
    response, 156f, 157
  receive pulse shaping, 160
  receiver, 161
  roll off factor, 157–158
  Shannon limit, 162–163
  16-QAM recovered
    constellation, 161f
  64-QAM recovered
    constellation, 161–162,
    162f
  64-QAM systems, 160
  square root raised cosine
    frequency response, 159f
    impulse response, 160f
  video carriers, 167
  using 0.25 roll off filter, 159
  zoomed in, 157, 157f
Real-time protocol (RTP), 173
Real-time streaming protocol
    (RTSP), 173
Resolutions, 5–6
RGB color model, 8, 8f
Roll off factor, 157
Routers, 170
RTP. *See* Real-time protocol

RTSP. *See* Real-time streaming
    protocol
Run length encoding, 106

**S**

S-video, 64
  cable, 65f
Salt-and-pepper noise, 141
  *See also* Mosquito noise
  filtering, 141
  high frequency, 143
  median filter, 142, 142f
    area, 143f
    forms, 142
  non-linear filtering, 141
  random noise median filter,
    142f
  sharp vertical edge, 142
Sampling, 11
  *See also* Aliasing; Nyquist
    sampling rule
  ADC, 11
  CCD, 11
  CMOS technology, 11
  color sampling, 12
  faster-moving signal, 13, 13f
  one-dimensional signal, 12
  slower-moving signal, 13, 13f
  spatial, 11
SAP. *See* Session
    announcement protocol
Scan-line duplication, 42, 43f,
    45
Scan-line interpolation, 42,
    43f, 45
SD. *See* Standard definition
SDH. *See* Synchronous digital
    hierarchy
SDI. *See* Serial data interface
Segmentation algorithm, 185
  contour integral engine,
    190f
  expanding circle template,
    189f
  expanding square template,
    186f, 188f
  FPGA implementation, 190
  template matching, 185–189

Sequence Parameter Set (SPS),
    139–140
Serial data interface (SDI), 61
  cable, 62f
  data rate, 62t
Session announcement
    protocol (SAP), 178
Session initiation protocol
    (SIP), 179
Setup time, 201, 202f
Shannon limit, 162–163
Signal bandwidth modulation
  *See also* Video demodulation;
    Video modulation
  high-frequency content, 152
  I and Q baseband signals,
    150–151
  using low-pass filter, 152
  orthogonal components, 150
  QPSK constellation, 150
  separated component plots,
    151f
  sharp transitions, 151
  signal frequency width
    modulation, 153f
  2D complex-constellation
    plane, 150
  two-dimensional signal,
    150–151
  up or down conversion, 152
Signal to noise power ratio
    (SNR), 77–78
Sinc impulse response, 153f
SIP. *See* Session initiation
    protocol
16-QAM
  constellation, 149f
  raised cosine filter, 161f
  video demodulation,
    148–149, 149f
  video modulation, 148–149,
    149f
64-QAM, 149
  raised cosine filter, 160
  recovered constellation,
    162f
  video demodulation, 149
  video modulation, 149
Slice, 117

Smooth streaming, 176
SNR. *See* Signal to noise power ratio
SONET. *See* Synchronous optical network
Spatial image analysis, 144
Spatial line doubling, 41
Spinning wheel experiment, 14
−1/8 revolution, 14–15, 16f
¼ of revolution, 14–15, 14f
½ of revolution, 14–15, 15f
1/8 of revolution, 14–15, 14f
$1^{1}/_{8}$ revolution, 14–15, 16f
7/8 of revolution, 14–15, 15f
counterclockwise $^{3}/_{4}$ of revolution, 14–15, 15f
eight times per second, 14–15, 15f
Sports coverage, 172
SPS. *See* Sequence Parameter Set
Standard definition (SD), 5
Sum of absolute differences (SAD). *See* Minimum absolute differences (MAD)
Synchronous digital hierarchy (SDH), 174
Synchronous optical network (SONET), 174
SystemConsole debugger, 204
SystemConsole host application, 208

**T**
Target, 204
TCP. *See* Transmission control protocol
TCP/IP. *See* Transmission control protocol over internet protocol
Temporal processing, 49
3:2 pull-down technique, 47
Timing analysis, 201–202
check meets timing design, 202–203
design fixing, 203–204

hold time, 202, 202f
meet timing, 202
propagation delay, 202
setup time, 201, 202f
Translucent, 50
Transmission control protocol (TCP), 172–173
Transmission control protocol over internet protocol (TCP/IP), 172–173
Transport protocols
error-free communication, 173
IP networks, 172
real-time protocol, 173
real-time streaming protocol, 173
streaming video and audio, 173
TCP/IP, 172–173
user datagram protocol, 173
Transport stream, 172
Transport-stream packets, 172
Triple frame buffer, 192
24 fps film, 47
Two-dimension (2-D)
complex-constellation plane, 150
FIR filters, 194
median filters, 194

**U**
UAV. *See* Unmanned Aerial Vehicle
UDP. *See* User datagram protocol
Underflow, 209
Uniform quantizing, 78
Unmanned Aerial Vehicle (UAV), 1–2
Up conversion, 152
User datagram protocol (UDP), 173

**V**
Variance technique, 184–185
VBV. *See* Video buffer verifier

VCEG. *See* Video Coding Experts Group
Vector quantization, 79
VGA, 64
Video, 1, 49
analog signal, 1
carriers, 167
channels, 152
content, 40
data, 141
decompression, 143
digital video processing, 1–2
frame, 29, 30f, 193
horizontal scan periods, 1
horizontal synchronization signal, 1
images, 67
over internet issues, 175–176
market sizes, 3
raster scan, 2f
targets, 2–3
Video buffer verifier (VBV), 121–122
Video Coding Experts Group (VCEG), 126
Video compression, 143
algorithms, 175
B-frame, 119
buffering, 120
decoder, 123–124, 123f
encoder, 122–123, 122f
frame processing order, 114–116
I-frame, 117–118, 117t
macroblock, 112, 113f
motion estimation, 114–116
P-frame, 118–119, 118t
quantization scale factor, 120
input decoder buffer, 121, 121f
Mquant, 120–121
VBV model, 121–122
rate control, 119
Video conferencing, 178
full-featured system, 178–179
H.320 standard, 178–179
holding long distance meetings, 179

session initiation protocol, 179
Video deinterlacing, 39–40
  bob deinterlacing, 41, 43
  cadence detection standard. *See* Cadence detection standard
  electron gun, 39
  interlaced video, 39, 40f, 41f
  interpolation, 41
  logic requirements
    DDR memory interface, 46
    deinterlacer design source, 46
    deinterlacers, 46
    external RAM bandwidth, 46
    formula, 46
    higher resolution image, 46
    memory, 46
    memory bandwidth, 46
    motion-adaptive deinterlacing, 46
    PAL video source, 46
    scan-line duplication algorithm, 45
    scan-line interpolation algorithm, 45
    weave deinterlacing method, 45
  motion-adaptive deinterlacing. *See* Motion-adaptive deinterlacing
  mouse teeth effect, 41, 42f
  progressive frame, 39
  progressive scanning, 40
  scan-line duplication, 42, 43f
  scan-line interpolation, 42, 43f
  spatial line doubling, 41
  video content, 40
  weave deinterlacing, 40–41, 42f, 43
Video demodulation, 147
  *See also* Video modulation; Signal bandwidth modulation
  baseband waveform, 147
  carrier frequency, 147

constellation, 147–148
constellation size and bit rate, 148t, 149t
constellation trajectory, 151f
frequency bandwidth, 150
I-Q plane, 147–148
line-of-sight systems, 150
non-line-of-sight system, 150
QPSK, 147
QPSK constellation, 148f
receiver, 150
transmitter, 150
16-QAM, 148–149, 149f
64-QAM, 149
Video interfaces, 61
  component video, 65
    cables, 65f
  CVBS, 64
    composite cable, 64f
  display port, 61–62, 62f
  DVI, 63
    connector, 63f
  HDMI, 63
    connector, 63f
  S-video, 64
    cable, 65f
  SDI, 61
    cable, 62f
    data rate, 62t
  VGA, 64
    monitor connector, 64f
Video modulation, 147
  *See also* Video demodulation; Signal bandwidth modulation
  baseband waveform, 147
  carrier frequency, 147
  constellation. *See* Constellation
  frequency bandwidth, 150
  I-Q plane, 147–148
  line-of-sight systems, 150
  non-line-of-sight system, 150
  QPSK, 147
  receiver, 150
  transmitter, 150
  16-QAM, 148–149, 149f
  64-QAM, 149

Video noise, 141
  mosquito noise. *See* mosquito noise
  salt-and-pepper noise. *See* Salt-and-pepper noise
Video over IP, 169
  encapsulation, 171
  internet protocol. *See* Internet protocol
  IP transport. *See* Internet protocol transport
  multicast video, 177–178
  transport protocols. *See* Transport protocols
  video conferencing. *See* Video conferencing
  video over internet issues, 175–176
  video streaming. *See* Video streaming
  video streams. *See* Video streams
Video processing, 191
  DDR memory, 191
  external memory bandwidth calculation, 194–199
  frame buffer. *See* Frame buffer
  motion-adaptive deinterlacer, 195
  on-chip memory calculation, 199–200
  requirements, 191
  performance, 5, 9–10
Video resizing. *See* Video scaling
Video rotation, 67f
  aircraft terms, 68
  arbitrary reference point substitution, 68
  using formula, 67–68
  interpolation, 68
  mathematical relationship, 67
  matrix form, 67–68
  using $3 \times 3$ matrix, 68
  three-dimensional rotation, 68f

Video rotation (*Continued*)
  video and imaging
      applications, 68
  video images, 67
Video scaling, 29, 38
  algorithms, 29
  aspect ratio, 36, 37f
    anamorphic widescreen, 38
    of HDTV, 36
    of standard TV, 36
  bicubic scaling, 32, 33f
  bilinear scaling, 31, 31f
  four-phase scaler, 31, 32f
  implementation, 33, 34f
    FPGA suppliers, 34
    GUI of an IP function, 35,
        36f
    IP functions, 35
    line buffer, 33
    2-D scaler, 33, 35f
  nearest neighbor method,
      30–31
  taps, 31

2 x 2 to 4 x 4 pixel image,
    30f
Video streaming, 176
  HTTP, 176–177
  multicast application,
      177
  standards, 176
  technology development, 177,
      177f
  video formatting and playing,
      176
  video source, 176
Video streams, 172
  decode timestamp,
      171–172
  elementary stream, 171
  PES, 171
  presentation timestamp,
      171–172
  program stream, 171–172
  transport stream, 172
  transport-stream packets,
      172

Video trace monitor
    component, 207, 207f
Vignette image, 109

**W**
WAN. *See* Wide area
    networking
WDR. *See* Wide dynamic range
Weapon sight. *See* Night-vision
    goggle (NVG)
Weave deinterlacing, 40–41,
    42f, 43–45
Wide area networking (WAN),
    174
Wide dynamic range (WDR),
    57

**X**
XP. *See* Extended Profile

**Y**
YCrCb color space, 9, 9f

Printed and bound by CPI Group (UK) Ltd, Croydon, CR0 4YY

03/10/2024

01040324-0001